Communications in Computer and Information Science 573

Commenced Publication in 2007
Founding and Former Series Editors:
Alfredo Cuzzocrea, Dominik Ślęzak, and Xiaokang Yang

More information about this series at http://www.springer.com/series/7899

Pavel Braslavski · Ilya Markov
Panos Pardalos · Yana Volkovich
Dmitry I. Ignatov · Sergei Koltsov
Olessia Koltsova (Eds.)

Information Retrieval

9th Russian Summer School, RuSSIR 2015
Saint Petersburg, Russia, August 24–28, 2015
Revised Selected Papers

 Springer

Editors

Pavel Braslavski
Ural Federal University
Yekaterinburg
Russia

Ilya Markov
University of Amsterdam
Amsterdam
The Netherlands

Panos Pardalos
University of Florida
Gainsville, FL
USA

Yana Volkovich
Eurecat
Barcelona
Spain

Dmitry I. Ignatov
National Research University Higher School
    of Economics
Moscow
Russia

Sergei Koltsov
National Research University Higher School
    of Economics
Saint Petersburg
Russia

Olessia Koltsova
National Research University Higher School
    of Economics
Saint Petersburg
Russia

ISSN 1865-0929   ISSN 1865-0937 (electronic)
Communications in Computer and Information Science
ISBN 978-3-319-41717-2   ISBN 978-3-319-41718-9 (eBook)
DOI 10.1007/978-3-319-41718-9

Library of Congress Control Number: 2016932329

Printed on acid-free paper

This Springer imprint is published by Springer Nature
The registered company is Springer International Publishing AG Switzerland

Preface

The 9th Russian Summer School in Information Retrieval (RuSSIR 2015) was held during August 24–28, 2015, in St. Petersburg, Russia.[1] The school was co-organized by the National Research University Higher School of Economics[2] and the Russian Information Retrieval Evaluation Seminar (ROMIP).[3]

The RuSSIR school series started in 2007 and has developed into a renowned academic event with extensive international participation. Previously, RuSSIR took place in Yekaterinburg, Taganrog, Petrozavodsk, Voronezh, St. Petersburg, Yaroslavl, Kazan, and Nizhny Novgorod. RuSSIR courses were taught by many prominent International Researchers in Information Retrieval and Related Areas.

The RuSSIR 2015 program featured courses focusing on social network analysis and graph mining along with traditional Information Retrieval topics. The program consisted of two invited lectures, eight courses running in two parallel sessions, three sponsor's talks, and the RuSSIR 2015 Young Scientist Conference.

The school welcomed 134 participants selected based on their applications. The majority of students came from Russia, but there were also 15 students from Europe and seven from the rest of the world. The RuSSIR audience comprised undergraduate, graduate, and doctoral students, as well as young academics and industrial developers. In total, 169 people participated including students, sponsor representatives, lecturers, and organizers.

The participation was free of charge thanks to the sponsors. In addition, 20 accommodation grants were awarded to Russian participants by the Higher School of Economics and 14 European-based students received travel support from the European Science Foundation (ESF)[4] through the ELIAS network.[5]

The main RuSSIR program was compiled based on peer-review of 15 submitted course proposals; five of them were selected for presentation. Additionally, there were five invited courses on the main topic of RuSSIR 2015, i.e., social network analysis and graph mining. Overall, the school program consisted of two plenary courses and eight regular courses run in two parallel tracks:

- Data Science for Massive (Dynamic) Networks, Panos M. Pardalos (University of Florida, USA)
- Community Detection in Networks, Santo Fortunato (Aalto University, Finland)
- Text Quantification, Fabrizio Sebastiani (Qatar Computing Research Institute, Qatar)

---

[1] http://romip.ru/russir2015/.

[2] http://www.hse.ru/en/.

[3] http://romip.ru/en/.

[4] http://www.esf.org/.

[5] http://www.elias-network.eu/.

- Leveraging Knowledge Graphs for Web Search, Gianluca Demartini (University of Sheffield, UK)
- Online/Offline Evaluation of Search Engines, Evangelos Kanoulas (University of Amsterdam, The Netherlands)
- Models of Random Graphs and Their Applications to the Web-graph Analysis, Andrey Raigorodsky (Moscow Institute of Physics and Technology, Moscow State University, and Yandex, Russia)
- Visual Object Recognition and Localization, Ivan Laptev (Inria Paris-Rocquencourt, France)
- Contextual Search and Exploration, Charles L.A. Clarke (University of Waterloo, Canada), Jaap Kamps (University of Amsterdam, The Netherlands), Julia Kiseleva (Eindhoven University of Technology, The Netherlands)
- Big Data-Driven Logistics, Athanasios Migdalas (Luleå University of Technology, Sweden)
- Big Data Analytics with R, Athanasia Karakitsiou (Luleå University of Technology, Sweden)

Sponsoring organizations made three scientific presentations in addition to the main school program. Eugene Kharitonov from Yandex presented a novel methodology that allows less exhaustive online experimentation for search engines; Vladimir Gulin from Mail.ru focused on user behavior analysis; Dmitry Bugaychenko from Ok.ru discussed the connection between the size of data and its usefulness.

For the ninth time the RuSSIR Young Scientist Conference was organized within the school program. The conference facilitated a scientific dialog between young researchers from different areas such as mathematics, computer science, and linguistics as well as social and media sciences. The conference ran over two consecutive evenings and consisted of two parts: oral presentations and poster sessions. There were two types of submissions: full papers that underwent a thorough reviewing process and short poster notes. Out of 17 submitted full papers, six were accepted for oral presentation at the conference and are published in the current volume. During the poster sessions all participants had an opportunity to discuss and exchange their research results and ideas. As in the previous years, the Young Scientist Conference was one of the main highlights of the school.

Charles Clarke, Jaap Kamps, and Julia Kiseleva organized a hackathon as a part of their tutorial. The hackathon was designed in a way similar to the TREC 2015 Contextual Suggestion Track.[6] The participants were asked to recommend to the lecturers a number of places to visit in St. Petersburg, based on the lecturers' profiles and external sources about the city. The hackathon attracted 30 students who formed ten teams. The winning teams were selected based on the originality, relevance, and efficiency of the proposed solutions:

- MAD IT (Best System Award) – Maria Zagulova, Andrey Poletaev, Dmitry Zhelonkin, Ivan Grechikhin, Tatiana Nikulina

---

[6] https://sites.google.com/site/treccontext/.

- SalsaRoulette (Best Presentation Award) – Navid Rekabsaz, Larisa Adamyan, Ioanna Miliou, Aldo Lipani
- Sleep_deprived (Most Original Approach Award) – Sagun Pai, Sheikh Muhammad Sarwar

The lecturers visited several of the recommended venues over the weekend and confirmed the high relevance of the suggestions.

RuSSIR 2015 was accompanied by two social events. The welcome party was held on the HSE premises on the first day of the school. It aimed to give participants and lecturers an opportunity to meet each other in an informal environment. Also, during the welcome party the RuSSIR sponsors had a chance to present their companies. A boat trip on the fourth day of the school took the participants along the beautiful Neva river and its numerous branches to show the magnificent view of the imperial St. Petersburg by night.

The volume features two sections: lecture notes ranging from 13 to 51 pages and six selected revised papers from the associated Young Scientist Conference (up to 20 pages each). The previous proceedings are published in the CCIS series as Vol. 505[7].

The 9th Russian Summer School in Information Retrieval was a successful event: It brought together participants with diverse backgrounds from Russia and abroad and facilitated a cross-disciplinary exchange of experience and ideas. Students had a unique opportunity to learn new material that is not usually present in university curricula and got feedback from peers and lecturers during the poster sessions and informal communications. The event contributed to supporting a lively information retrievel community in Russia and establishing ties with international colleagues. The organizers received very positive feedback from attendees on all aspects of the school.

We would like to thank all the local Organizing Committee members (especially, Daria Yudenkova) for their commitment, which made the school possible, all the Program Committee members for their time and efforts ensuring a high level of quality for the RuSSIR 2015 program, and, in particular, all the lecturers and students who came to St. Petersburg and made the school such a success. We also would like to thank the student volunteers who contributed to the school organization on-site. Our special gratitude goes to Maxim Gubin, who was responsible for legal and financial matters.

We appreciate the generous financial support from our sponsors: National Research University Higher School of Economics (main organizer); golden sponsors: Yandex,[8] Mail.Ru,[9] and the ELIAS network[10] of the European Science Foundation; bronze sponsors: Google,[11] Rambler,[12] JetBRAINS,[13] and the Russian Foundation for Basic

---

[7] http://dx.doi.org/10.1007/978-3-319-25485-2.

[8] http://yandex.com.

[9] http://go.mail.ru/.

[10] http://www.elias-network.eu/.

[11] http://google.com/.

[12] http://rambler-co.ru/en/.

[13] http://www.jetbrains.com/.

Research[14]; Yana Volkovich was supported by a special travel grant[15]. We are also grateful to Springer, namely, Alfred Hofmann and Aliaksandr Birukou, for their support.

May 2016

Pavel Braslavski
Ilya Markov
Panos Pardalos
Yana Volkovich
Dmitry I. Ignatov
Sergei Koltsov
Olessia Koltsova

---

[14] http://www.rfbr.ru/rffi/eng.

[15] The People Programme (Marie Curie Actions, from the FP7/2007- 2013) under grant agreement no. 600388 managed by REA and ACCIO.

Organization

The conference was organized by the National Research University Higher School of Economics (Saint Petersburg, Russia) and Internet Studies Laboratory[1].

Program Committee

Ilya Markov (Co-chair)	University of Amsterdam, The Netherlands
Panos Pardalos (Co-chair)	University of Florida, USA
Ismail Sengor Altingovde	Middle East Technical University, Turkey
Pavel Braslavski	Ural Federal University/Kontur Labs, Russia
Ingo Frommholz	University of Bedfordshire, UK
Max Gubin	Facebook, USA
Djoerd Hiemstra	University of Twente, The Netherlands
Katja Hofmann	Microsoft Research, UK
Dmitry Ignatov	National Research University Higher School of Economics, Russia
Hideo Joho	University of Tsukuba, Japan
Sergei Koltsov	National Research University Higher School of Economics, Russia
Daan Odijk	University of Amsterdam, The Netherlands
Panos Pardalos	Industrial and Systems Engineering University of Florida, USA
Paolo Rosso	Technical University of Valencia, Spain
Stefan Rueger	Open University, UK
Anne Schuth	University of Amsterdam, The Netherlands
Pavel Serdyukov	Yandex, Russia
Natalia Vassilieva	HP Labs, Russia
Svitlana Volkova	Johns Hopkins University, USA
Ingmar Weber	Qatar Computing Research Institute, Qatar

Young Scientist Conference Program Committee

Yana Volkovich (Chair)	Eurecat, Spain
Ismail Sengor Altingovde	Middle East Technical University, Turkey
Pavel Braslavski	Ural Federal University/Kontur Labs, Russia
Santo Fortunato	Aalto University, Finland
David Gleich	Purdue University, USA
Dmitry Ignatov	National Research University Higher School of Economics, Russia

---

[1] https://linis.hse.ru/en/.

Vladimir Ivanov	Kazan Federal University, Russia
Evangelos Kanoulas	University of Amsterdam, The Netherlands
Athanasia Karakitsiou	Lulea University of Technology, Sweden
Julia Kiseleva	Eindhoven Technical University, Russia
Sergei Koltsov	National Research University Higher School of Economics, Russia
Olessia Koltsova	National Research University Higher School of Economics, Russia
Natalia Konstantinova	University of Wolverhampton, UK
Ilya Markov	University of Amsterdam, The Netherlands
Marc Miquel	Amical Viquipèdia, Spain
Minsu Park	Cornell University, USA
Andrei Raigorodskii	Moscow State University, Russia
Fabrizio Sebastiani	Qatar Computing Research Institute, Qatar
Pavel Serdyukov	Yandex, Russia
Julia Stoyanovich	Drexel University, USA
Denis Turdakov	Institute for System Programming of RAS, Russia
Natalia Vassilieva	HP Labs, Russia
Svitlana Volkova	Johns Hopkins University, USA
Ingmar Weber	Qatar Computing Research Institute, Russia
Robert West	Stanford University, USA

Proceedings Chair

Dmitry Ignatov	National Research University Higher School of Economics, Russia

Organizing Committee

Sergei Koltsov (Chair)	National Research University Higher School of Economics, Russia
Olessia Koltsova	National Research University Higher School of Economics, Russia
Dmitry Ignatov	National Research University Higher School of Economics, Russia
Elena Gruzinskaya	National Research University Higher School of Economics, Russia
Daria Yudenkova	National Research University Higher School of Economics, Russia

Young Scientist Conference Organizing Committee

Dmitry Ignatov	National Research University Higher School of Economics, Russia
Sergey Nikolenko	National Research University Higher School of Economics, Russia
Lidia Pivovarova	University of Helsinki, Finland

Steering Committee

Pavel Braslavski	Ural Federal University/Kontur Labs, Russia
Dmitry Chalyy	Yaroslavl Demidov State University, Russia
Alexander Goncharov	CVisionLab, Russia
Maxim Gubin	Google, USA
Nikolay Karpov	National Research University Higher School of Economics, Russia
Ksenia Rogova	Katholieke Universiteit Leuven, Belgium
Alexander Sychev	Voronezh State University, Russia
Natalia Vassilieva	HP Labs, Russia
Nikita Zhiltsov	Kazan Federal University, Russia

Partners and Sponsoring Institutions

Gold sponsors
Yandex
Mail.ru
ELIAS Network
Bronze sponsors
Google
Rambler
Jet Brains
Russian Foundation for Basic Research

Tutorial Abstracts

Community Detection in Networks

Santo Fortunato

Aalto University, Espoo, Finland
santo.fortunato@aalto.fi

Abstract. The course is focused on one of the most popular topics in the network science: detection of communities in networks. Communities are usually conceived as subgraphs of a network, with a high density of links within the subgraphs and a comparatively lower density between them. I introduce the elements of the problem, e.g. definitions of community and partition, and dwelve into some of the most popular methods. Special attention is devoted to the optimization of global quality functions, like Newmna-Girvan modularity, and to their limits. Finally we discuss the crucial issue of testing, both on artificial benchmark graphs with built-in community structure and on real networks.

Keywords: Network science · Community detection

Reference

1. Fortunato, S.: Community detection in graphs. Phys. Rep. **486**(35), 75–174 (2010)

Visual Object Recognition and Localization

Ivan Laptev

INRIA Paris-Rocquencourt, Paris, France
ivan.laptev@inria.fr

Abstract. The goal of this course was to introduce state-of-the-art methods for large scale image recognition and retrieval. The course contained lectures and one practical session. The lectures covered recent image representations for object recognition (HOG [1], SIFT [2], DPM [3], BOF [4–7]) as well as modern machine learning techniques (SVM [8], CNN/Deep Learning [9–11]). Besides lectures, the course included a guided practical session where students were able to implement basic techniques for object recognition. As a result of the course, participants have learned about techniques enabling efficient search of particular object instances among billions of images. The participants have also learned about most recent advances in Deep Learning enabling close-to-human performance for such tasks as face recognition and object category recognition.

Keywords: Computer vision · Image recognition · Image representation · Object localization

References

1. Dalal, N., Triggs, B.: Histogram of oriented gradients for human detection (2005)
2. Lowe, D.: Distinctive image features from scale-invariant keypoints **60**(2), 91–110 (2004)
3. Felzenszwalb, P., Girshick, R., McAllester, D., Ramanan, D.: Object detection with discriminatively trained part-based models. IEEE PAMI **32**(9), 1627–1645 (2010)
4. Csurka, G., Dance, C., Fan, L., Willamowski, J., Bray, C.: Visual categorization with bags of keypoints. In: ECCV Workshop (2004)
5. Sivic, J., Zisserman, A.: Video Google: a text retrieval approach to object matching in videos (2003)
6. Perronnin, F., Sánchez, J., Mensink, T.: Improving the fisher kernel for large-scale image classification. In: Daniilidis, K., Maragos, P., Paragios, N. (eds.) ECCV 2010. LNCS, vol. 6314, pp. 143–156. Springer, Heidelberg (2010)
7. Cinbis, R., Verbeek, J., Schmid, C.: Segmentation driven object detection with fisher vectors. In: ICCV. IEEE (2013)
8. Vapnik, V.: Statistical Learning Theory. Wiley, NY (1998)
9. LeCun, Y., Bengio, Y.: Convolutional networks for images, speech, and time-series. In: Arbib, M.A. (ed.) The Handbook of Brain Theory and Neural Networks. MIT Press (1995)
10. Krizhevsky, A., Sutskever, I., Hinton, G.: Imagenet classification with deep convolutional neural networks. In: NIPS (2012)
11. Girshick, R., Donahue, J., Darrell, T., Malik, J.: Rich feature hierarchies for accurate object detection and semantic segmentation. In: CVPR, pp. 580–587 (2014)

Text Quantification

Fabrizio Sebastiani

Qatar Computing Research Institute, Doha, Qatar
fsebastiani@qf.org.qa

Abstract. In a number of applications involving text classification in recent years it has been pointed out that the final goal is not determining which class (or classes) individual unlabeled documents belong to, but determining the prevalence (or "relative frequency") of each class in the unlabeled data. The latter task is known as text quantification (or prevalence estimation, or class prior estimation). The goal of this course was to introduce the audience to the problem of quantification, techniques that have been proposed for solving it, metrics used to evaluate them, applications in fields such as information retrieval, machine learning, and data mining, and to the open problems in the area.

Keywords: Text classification · Text quantification

Reference

1. Sebastiani, F.: Machine learning in automated text categorization. ACM Comput. Surv. **34**(1), 1–47 (2002)

Contents

Tutorial Papers

Contextual Search and Exploration

Julia Kiseleva[1]([✉]), Jaap Kamps[2], and Charles L.A. Clarke[3]

[1] Eindhoven University of Technology, Eindhoven, The Netherlands
julianakiseleva@gmail.com
[2] University of Amsterdam, Amsterdam, The Netherlands
[3] University of Waterloo, Waterloo, Canada

Abstract. Personalized (mobile) devices are radically changing information access tools, with rich context allowing for far more powerful, personalized search. Rather than retrieving a "document" on the topic of a "query," the rich contextual information allows for tailored search and recommendation, and solve user's complex tasks by taking into account complex constraints, exploring options, and combining individual answers into a coherent whole. This paper reports on a RuSSIR 2015 course covering the challenges of contextual search and recommendation, with a concrete focus on the venue recommendation task as run as part of TREC 2012–2015. It consisted of both lectures and hands-on "hackathon" sessions with data derived from the TREC task.

1 Introduction

The ubiquitous availability of information on the web and personalized (mobile) devices has a revolutionary impact on modern information access, challenging both research and industrial practice. Searchers with a complex information need typically slice-and-dice their problem into several queries and subqueries, and laboriously combine the answers post hoc to solve their tasks. Rich context allows for far more powerful, personalized search, without the need for users to write long complex queries. Consider planning a social event at the last day of RuSSIR, in the unknown city of Saint Petersburg, factoring in distances, timing, and preferences on budget, cuisine, and entertainment. Rich context and profiles in combination with a curated set of web data allow us to solve complex tasks with just a simple query: `entertain me`. Rather than retrieving a "document" on the topic of a "query," the rich contextual information allows for tailored search and recommendation, and solve their complex task by taking into account complex constraints, exploring options, and combining individual answers into a coherent whole.

This RuSSIR 2015 course covered the challenges of contextual search and recommendation, with a concrete focus on the venue recommendation task as run as part of TREC 2012–2015 Contextual Suggestion Track. It consisted of both lectures and hands-on "hackathon" sessions with data derived from the TREC task. Our goal was enabling students to understand the challenges and opportunities of contextualized search over entities, and learn effective approaches for

P. Braslavski et al. (Eds.): RuSSIR 2015, CCIS 573, pp. 3–23, 2016.
DOI: 10.1007/978-3-319-41718-9_1

the concrete application to venue recommendation domain, as well as obtain hands-on experience with developing and evaluating personalized search and recommendation approaches.

The rest of this paper is structured as follows. After this introduction, Sect. 2 gives an overview of approaches to contextual search and exploration, focusing on venue recommendation. Next, Sect. 3 details how to set up an experiment to evaluate contextual suggestion based on the TREC track. Section 4 provides detailed approaches of to using contextual information in modeling search and interaction behavior. Then, Sect. 5 discusses the setup and results of the hackathon. Finally, Sect. 6 concludes the paper with some discussion on the outcome of the lectures and hackathon.

2   Approaches to Contextual Search and Exploration

In the first session, Jaap provided an overview of the tutorial and hackathon, and introduced various approaches to contextual search and exploration. It is motivated by complex search tasks now requiring several independent searches and put the onus on the user to manage the overall task progress, and combine individual results into a coherent whole.

As explained before, the official goal of the course was to enable students to understand the challenges and opportunities of contextualized search over entities, and learn effective approaches for the concrete application to venue recommendation domain. The unofficial goal, however, was to have the students plan our time in St. Petersburg. The lecturers wanted to visit an amazing city but are clueless about what to do, and invented a course so the students attending RuSSIR will be planning our holiday in St. Petersburg! A special edition of the TREC Contextual Suggestion Track's batch task was run as a hackathon, with profiles of Charlie, Julia and Jaap, and 102 candidate venues in St. Petersburg (Palaces, Museums, Restaurants, Bars, Clubs, etc.) to visit. We asked the students to build a system that gives us the best venues to visit after the lectures.

2.1   Slogan #1: Standard IR Fails for Venue Recommendation!

The course focused on contextual search and exploration, with a planning problem as leading example. The overall goal is to address complex information needs on mobile devices, using rich contextual information and user profiles, and taking into account complex constraints, exploring options, and combining individual answers into a coherent whole. The specific focus is on venue or point of interest (POI) recommendation for travelers, e.g., Canadian and Dutch people in St. Petersburg in August. What are we going to do this evening? How to plan what to do in an unknown city? What to see? Where to eat? Where to drink? The most popular things? Or those that I like best? Is there actually a ballet performance tonight? How do I get from venue 1 to venue 2 to ...?

The venue recommendation problem gets as input: (1) a *start signal* such as an App click or generic query; (2) a *context* such as a location or city; and

(3) a *profile* of the user, containing explicit profile information such as age and gender, and implicit profile information such as likes/dislikes in other cities that can be derived from earlier interactions on a phone.

There are many travel sites online, including Tripadvisor, Foursquare, Yelp, Google Places, Yandex Cities, etc. Most of these sites offer venue search with some level of support for the context (typically the data is organized by location equated by city or country/region), and essentially no support for the profile (typically very limited personalization/customization).

Standard search is not getting us very far: there is no query or statement of request in the traditional sense, and just using a city name or venue type as query leads to very poor results, unless the context and profile are taken into account. This leads to our first slogan: *Standard IR Fails for Venue Recommendation!*.

2.2 Slogan #2: Location Is Context

Venue recommendation isn't the same as geographical or location based search. Geo search exists for a few decades within IR. It is typically using a selection of typical search engine queries, focusing on those queries where part of the query is, or has, a location. For example, think of a query like "`restaurant in beijing china`." Each of the queries tends to have an exact answer, which is the same for anyone issuing the query, e.g., the query "`taj mahal`" linking to http://www.tajmahal.gov.in/.

Benchmarks on Geo search include the Geo IR tracks at CLEF[1] and the Geo-Time task at NTCIR.[2] Approaches to Geo IR typically use special resources or knowledge bases, with explicit locations like cities and countries, or POIs with GPS coordinates. The task is mostly about identifying the location part of the query, and mapping it to these resources, and search engines provide APIs for this.

Venue recommendation is different from Geo search. In venue recommendation, the query is a normal generic query without a location, e.g., "`restaurant`," "`bar`," or "`museum`." But the result should take the location into account: location is the *context* of the request, and venues too far will never be relevant. So a different context means an entirely different result set. This leads to our second slogan: *Location is Context*.

2.3 Slogan #3: Need to Blend Search and Recommendation

Venue recommendation isn't the same as collaborative recommendation. Work on recommendation is dominated by collaborative filtering. Here the input is a large set of ratings by many people, and the profile of a person x. The output is a ranked list of items y unrated by x, that x will rate high, based on people similar to x giving high ratings to y. There are many approaches to recommendation, the standard collaborative filtering approach treats each person as a vector of

---

[1] GeoCLEF 2005–2008, see: http://www.clef-initiative.eu/track/geoclef.
[2] NTCIR 8–9, 2010–2011, see: http://metadata.berkeley.edu/NTCIR-GeoTime/.

ratings: like/dislike/unknown, and looks at similar persons by cosine similarity over these vectors. The person most similar to person x is x herself, so we pick the next most similar person. Clustering and machine learning approaches are used to learn patterns in the training data, and to make predictions on unseen data.

Venue recommendation is different from collaborative recommendation. Collaborative recommendation assumes rich profiles of many users, but suffers from cold start problems: new users without history, and very sparse profiles. Most e-commerce providers see their users a few times a year, and have a continuous cold start problem. Hence we need to factor in search or content based recommendation.

We need aspects from both search and recommendation. Venue recommendation is not just serendipitous recommendation, such as a random book you like, but focused on a specific information need. But there is also no explicit query to match, such as when querying to look up the location of a particular known venue, but it is initiated by a generic query ("`st. petersburg`") or App click. This leads to our third slogan: *Need to Blend Search and Recommendation.*

2.4   Slogan #4: Search Is Getting Personal

There is no one size fits all approach to venue recommendation. Contextual search and recommendation requires a radical departure from the query-response paradigm of prototypical search, which takes as input a short query, and outputs a ranked list of results. This approach is still dominating research and industrial practice, with current systems excelling at short narrow scoped queries, heavily optimized against log data.

In terms of user satisfaction and user experience, this is likely a local optimum, where we cannot break out without changing something more fundamental. This implies that we need to step away from this "ten blue links" approach, and think about ways to support the user's whole search task. Currently, the emerging intelligent personal assistants come closest to this: Google Now, Microsoft Cortana, Apple's Siri, Facebook's M, etc. are starting a new search paradigm where context and profile information is key.

Your phone knows you, may know more about you than you know yourself: your work moved online into the clouds, and your personal life moved as well— everything you ever did is there... This data is personal, but also highly curated with clear entities and structure, allowing for powerful graph search with highly expressive queries.

We need to go beyond the query-response paradigm. This is not about personalization in terms of slightly changing the ranking by swapping some results, but an extreme form of personalization where different users get fundamentally different results: the profile is determines your result set—*you* are the query. This leads to our fourth slogan: *Search is Getting Personal.*

2.5   Wrap Up

We discussed venue recommendation as a personalized and contextualized task with complex constraints. Location is only part of the problem: it is not the same as geographical search. Profiles matter but are sparse: it not the same as collaborative recommendation. It is a form of extreme personalization: it cannot be handled by a one-size-fits-all approach. Sessions are highly interactive complex search going beyond the traditional query-response paradigm.

In the next section, we discuss a simplified form of venue recommendation for which a benchmark evaluation is being developed at TREC.

3   The TREC Contextual Suggestion Track

In the second 90-min session, Charlie provided an overview of the TREC Contextual Suggestion Track[3], which creates open data collections for evaluating contextal suggestion and point-of-interest recommendation. Since 2012 [9,10,14], the Contextual Suggestion Track has operated as part of the TREC[4] series of evaluation experiments, sponsored by the U.S. National Institute of Standards and Technology. The track imagines a traveler in a unknown city seeking sites to see and things to do that reflect his or her own personal interests, as inferred from their interests in their home city. For example, a group of information retrieval researchers visiting Saint Petersburg in August, such as the authors of this tutorial, should be directed to museums, restaurants, and bars that reflect their individual tastes. According to the Second Strategic Workshop on Information Retrieval in Lorne [4]:*"Future information retrieval systems must anticipate user needs and respond with information appropriate to the current context without the user having to enter an explicit query..."* The TREC Contextual Suggestion Track establishes an evaluation framework allowing researchers to investigate this problem, at least within the limited domain of point-of-interest recommendation.

The tutorial session began with an overview of task as it operated from 2012 to 2014 [9,10,14]. As input to the task, participating research groups were given a set of profiles, a set of example suggestions, and a set of contexts. Each profile corresponded to a single user, indicating that users preference with respect to each example suggestion, while each context represented a target city that the user might visit. For each profile/context pairing, participating researchers were required to return a ranked list of 50 proposed suggestions. Each suggestion was expected to be appropriate to the profile (based on the user's preferences) and the context (according to the target city). Profiles correspond to the stated preferences of real individuals, primarily recruited through crowdsourcing. These crowdsourced workers first judged example attractions in seed locations, representing their home cities, and later returned to judge suggestions proposed by the participating research groups for various target cities.

---

[3] See: http://sites.google.com/site/treccontext/.

[4] See: http://trec.nist.gov/.

Most of this overview was drawn from track reports, which can be consulted for detailed information [9, 10, 14]. In the remainder of the session we discussed a number of issues related to the structure of the track, as detailed below, as well as the lessons learned from it. The tutorial ended with a discussion of ongoing and future work.

3.1   Issue #1: Assessor Quality

The first of these issues concerns the quality of assessment provided by crowd-sourced workers, who are not real travelers. Can we assume that these workers will provide judgments that accurately and consistently reflect their own opinions? We discussed ways in which worker/assessor quality can be measured, given that the degree to which an assessor likes or dislikes a point-of-interest attraction is purely a subjective question. We cannot simply look at agreement between assessors to determine assessment quality, as we would do for a traditional TREC retrieval task.

Each assessor has the implicit goal of ordering the systems according to their true ranking Thus, we measure assessor consistency by comparing the system ranking implied by the judgments of a single assessor with the average system ranking implied by the judgments of all assessors. In the tutorial, we examined the results of studying assessor consistency over TREC 2013 results [11].

The goal of the study was the identification of careful and consistent assessors in the early stages of the experiment, allowing us to minimize assessment costs and improve assessment quality. Unfortunately, while consistency can be high for some assessors, and appears reasonable for most assessors, we were unable to find a method of reliably detecting assessors. Moreover, assessors themselves do not remain consistent from context to context. However, despite this lack of consistency on the part of individual assessors, the group as a whole were able to identify significant differences between systems. Moreover, other research [12] into selecting the number of assessors to employ, supports the numbers of assessors selected for the TREC tasks.

3.2   Issue #2: Limitations of Evaluation Measures

The TREC Contextual Suggestions Tracks use precision@5 and mean reciprocal rank (MRR) as their primary evaluation measures. Unfortunately, precision@5 implicitly assumes that a user will always look at exactly the first five results, no more and no less, while MRR implicitly assumes that the user stops at the first useful result. Can we create an evaluation measure that better matches user behaviour?

A user's reaction to a suggestions could be negative ("dislike"), as well as positive ("like") or neutral, and too many disliked suggestions may cause the user to abandon the results. On the other hand, by reviewing caption descriptions, the user may be able to quickly skip suggestions that are not of interest, reaching much deeper into the list than the first five. Building on the *time-biased gain* (TBG) framework of Smucker and Clarke [32], which recognizes time as a critical element in user modeling for evaluation, we developed an evaluation measure that directly accommodates these factors [13].

The tutorial presented this version of TBG, which is tailored to the Contextual Suggestion task, along with some motivation and results. This version of TBG accounts for the impact of descriptions and disliked suggestions, both of which are ignored by the official track measures. The measure models a user working their way through a ranked list of suggestions, pausing to investigate the webpages associated with descriptions they like. Gain—or benefit to the user—is recognized after the user views a page they like. Disliked suggestions may cause the user to stop browsing. The model has four parameters, reflecting the probabilities of taking certain actions and the time needed to take these actions. These parameters may be set through studies of actual user behaviour, as captured in query logs and other sources.

3.3   Issue #3: Reusability and Repeatability

One goal of the TREC Contextual Suggestion Track is the creation of reusable test collections for future experiments. Output from the TREC Tracks during 2012–2014 included judgments from hundreds of assessors for hundreds of suggestions across hundreds of cities. Can these suggestions and judgments be re-used to evaluate future system?

We are still working on this issue [17]. Unfortunately, the reusability of collections developed for TREC 2012–2014 has proven to be limited. One problem in these years is that each participating group developed their own sets of candidate attractions for each venue, as well as their own descriptions for these attractions. For TREC 2015 (see below) suggestions must be made from a closed set of attractions, which may improve reusability.

3.4   TREC 2015 and Beyond

The track continued for TREC 2015, but with a very different character. This year, we took a "living labs" approach. Participants provided a continuously running online engine, and our server connected crowdsourced users with suggestions provided by these engines. In addition, suggestions were limited to a pre-defined set, with the goal of improving reusability.

If the track continues into the future, we hope to transition to a continuously running online evaluation service, managing a federation of online recommendation engines. Ideally, the service could be used for evaluation experiments outside the bounds of TREC, perhaps with real travelers slowly replacing crowdsourced workers. We are looking for volunteers to help make these ideas work!

In the next section, we discuss detailed approaches of to using contextual information in modeling search and interaction behavior.

4   Using Contextual Information to Understand Searching and Browsing Behavior

In the fourth session,[5] Julia detailed approaches to use contextual information to model search and browsing behavior. Modern search still relies on the query-response paradigm, which is characterized by a sharp contrast between the richness of data in the index, and the relative poverty of information in the query, usually expressed in a few keywords to capture a complex need. This is particularly true in online search services, where the same query may be observed from many users, with considerable variations in their search intents. Contextual information is the obvious route to try to restore the balance, and behavioral data related to user's searching and browsing activities provides new opportunities to model contextual aspects of user needs.

The importance of contextual information in search applications has been recognised by researchers and practitioners in many disciplines, including recommendation systems, information retrieval, ubiquitous and mobile computing, and marketing. Context-aware systems [20,21] adapt to users operations and thus aim at improving the usability and effectiveness by taking context into account. In this work we consider two types of behavior: (1) 'searching'—when users are issuing queries and we are trying to improve search results (SERP) taking context of sessions into account; and (2) 'browsing'—when users are surfing a website and we are predicting their movements utilizing context.

The main research problem we are investigating is the value of context in searching and browsing user behaviour on web: *how to discover, model and utilize contextual information in order to understand and improve users' searching and browsing behaviour on web?* We start by giving an overview of context as used in the literature (in Sect. 4.1). We continue by developing a general analytic framework that views context aware search from the system's perspective (in Sect. 4.2). This analytic part defines a general framework for modeling context, and introduces the notions of optimal contextual models and useful contextual models. Next, we look at the impact of specific contextual aspects, starting with geographic location as static contextual aspect (in Sect. 4.3), and similar behavioral trails of search and browse actions as dynamic contextual aspects (in Sect. 4.4). Finally, we look at behavioral dynamics—changes in aggregated user behavioral features over time—such as the frequency of query revisions and SAT/DSAT clicks to detect changes in user satisfaction and drifts in query intent (in Sect. 4.5).

4.1   What Is (Not) Context?

In this section, we give a short overview of "context" in the literature. First, we give a broad overview of context as used in various field. Second, we detail

---

[5] The third session introduced the hackathon and the tools and data available for it, and will be discussed together with the outcome of the hackathon in the next section.

Table 1. The evolution of context definition

Context	Year
Location	1992
Taxonomy of explicit context	1999
Predictive features versus contextual	2002
Hidden context: clustering, mixture models	2004
Contextual bandits	2007
History of previous interaction	2008
Independence of predicted class	2011
Two level prediction model	2012
Focus on context discovery	2012–

the use of context in search systems. Third, we discuss the use of context in recommender systems.

Many interpretations of the notion of context have emerged in various fields of research like psychology, philosophy, and computer science [6]. In literature, a context was presented as additional (situational) information: a user's location [1], helping to identify people near the user and objects around [19], current date, season, and weather [7]. Later, the user's emotional status was added to the context-aware application, Dey et al. [15] broadened the definition to "any information that can characterise and is relevant to the interaction between a user and an application." These works typically assume that context is explicit and given by a domain expert, whereas our focus is on implicit contextual information.

In machine learning, context was considered as *contextual features* in supervised concept learning [35]. The contextual features are useful for classification only when they are considered in combination with other features. For example, in medical diagnosis problems, the patent's gender, age, and weight are often available. These features are contextual, since they (typically) do not influence the diagnosis when they are considered in isolation. Later it was discovered that a context may not necessarily be present in form of a single variable in the feature space. It can be hidden in the data. Turney [34] formulated the problem of recovering implicit context information and proposed two techniques: input data clustering and time sequence. According to Prahalad [29], a context has temporal (when to deliver), spatial (where), and technological (how) dimensions. In terms of interactive systems, Palmisano et al. [28] has shown that it was useful to consider the history of user interaction (changes in these entities). In Zliobaite [39] a context was defined as an artifact in the data that does not directly predict the class label, e.g. accent in speech recognition. Zliobaite et al. [40] proposed context-aware systems as two level prediction models for food sales. The timeline of the main milestones related to the research of context in predictive modeling is presented in Table 1.

In information retrieval, context of a search query often provides a search engine with meaningful hints for answering the current query better and can be utilised for ranking. Given a query, a search engine returns the matched documents in a ranked list to meet the user's information need. Understanding users' search intent expressed through their search queries is crucial to Web search. A web query classification has been widely studied for this purpose. Cao et al. [8] incorporates context information into the problem of query classification by using conditional random fields models (context is used to expand a feature space). This approach uses neighbouring queries and their corresponding clicked Web pages in search sessions as a context. Context-aware search adapts search results to individual search needs using contexts. While personalised search considers individual users long and/or short histories, context-aware search focuses on short histories of all users. Xiang et al. [37] adopts a learning-to-rank approach and integrates the ranking principles into a state-of-the-art ranking model by encoding the context as a feature of the model. The experimental results clearly show that this context-aware ranking approach improves the ranking of a commercial search engine.

In recommender systems, Adomavicius and Tuzhilin [2] showed that the *situation* in which a choice is made is important information. E.g., using a temporal context in a travel recommender system would provide a vacation recommendation in the winter that can be very different from the one in the summer. Similarly, in the case of personalised content delivery on a Web site, it is important to determine what content needs to be recommended to a customer. The purchase intent of a customer is considered as contextual information in an e-commerce application because different purchasing intents may lead to different types of behavior [2]. The purchase intent usually is considered as a hidden context which has to be derived. Then it can be used to select 'right' model. The context-aware recommenders utilize the information about a situation to make predictions. Palmisano et al. [28] defined a hierarchy of a context in the recommendation system they used the obtained contextual features to expand feature space. The other effective method for a context-aware rating prediction is Multiverse Recommendation based on the Tucker tensor factorization model [33]. Stern et al. [33] presented probabilistic model for generating personalised recommendations of items to users of a web service. Their system makes use of explicit context information in the form of a user (e.g. age and gender) and meta data of an item (e.g. author and manufacturer) in combination with collaborative filtering information from previous user behaviour in order to predict the value of an item for a user. The contextual information is integrated into the prediction process using a feature set expansion manner to produce the better recommendations. Rendle et al. [30] proposed a novel approach applying Factorization Machines to model contextual information and to provide context-aware rating predictions using context explicitly specified by a user to the set of predictive features.

4.2   General Definition of Useful Context

We will now discuss how to define a general analytical framework for context-aware systems. By defining a general framework, we can clarify concepts, and define the abstract problem underlying the use of context in concrete applications.

First, we define a general view of what is contextual information. Then we introduce how contextual information might be utilized. Let $\Theta = C_1 \times \cdots \times C_i \times \cdots \times C_N$ be the space of all possible contextual features associated with every data instance, where each C_i is a context. Denote $\theta_s \in \Theta$ as the contextual feature vector associated with each data instance s. Let $M : \Theta \times D \mapsto V$ be a predictive contextual model that maps each test instance $s \in D$ associated with the contextual information θ_s to a decision space V. Let $F(s, M(\theta_s, s)) : D \times V \mapsto \mathbb{R}$ be a function evaluates how good a model is. For example, in the case of the next action prediction, it foretells a next users' activity. The space of users' activities is the following set: $\{Search = a, Click\,on\,Ad\,Banner = b, Click\,on\,Recommendation = c\}$. In this case, our decision space V is the same as our data instance space D. An example of the evaluation function might be the number true predictions made by M over the test instance s. For instance, assume that the model M predicts the following set of activities $s = ababc$ as $M(\theta_s, s) = \underline{ab}e\underline{dc}$ then it makes three true predictions corresponding to the underlined activities, i.e. $F(s, M(\theta_s, s)) = 3$.

Let $T \subseteq D$ be a set of test instances and denote $Pr(s)$ as the probability that $s \in T$. The expectation of an evaluation function $F(s, M(\theta_s, s))$ over our test set is defined as $E[T, M] = \sum_{s \in T} Pr(s) * F(s, M(\theta_s, s))$. The value of the expectation $E[T, M]$ can be considered as an objective function that needs to be maximized. We assume that $\exists M^*$ which is a (sub-)optimal model, i.e. $M^* = \arg\max_M E[T, M]$. Essentially, the optimal model uncovers the optimal weights of each contextual feature (either static as location, or dynamic as search trail characteristics) in order to predict the outcome (such as the next action, or a result click, or a query revision).

Let C be a context with n categories: $C = \{c_1, \ldots, c_j, \ldots, c_n\}$ associated with each data instance $s \in D$. A context may have different categories, e.g. the geographical context can be divided into four categories such as continents: Europe, Africa, American, or Asia. For simplifying our discussion, we consider the context that have only two categories, as the discussion for the general case which includes than two categories is very similar. Assume that we have a context C with two categories c_1 and c_2 dividing the test set into two disjoint subsets T_1 and T_2 such that $T = T_1 \cup T_2$. Denote M_1 and M_2 as two predictive models built for the category c_1 and c_2 respectively. Let $P(c_1)$ and $P(c_2)$ are probabilities that a test instance belonging to the category c_1 and c_2 respectively.

Theorem 1. (Contextual Principle). *Let M^* be an (sub-)optimal model for T then it is a combination of M_1^* and M_2^*. Where M_1^* is an optimal model for T_1 and M_2^* is an optimal model for T_2.*

Theorem 1 (the formal proof is provided in [26]) shows that the problem of finding the best model for every test instance can be solved by considering the sub-problems of finding optimal models for test instances in each individual contextual category. This is a technical result of a desirable property that allows us to work on customization to user types or profiles, or personas, rather than personalization to specific individuals.

Nevertheless, in practice finding an optimal model for each contextual category is usually as hard as finding an optimal model for the whole data. Indeed, it is usually the case that the type of model is chosen in advance, e.g. Markov models. Model's parameters are estimated from training data D. Under this circumstance, contextual predictive analytics seeks for a context such that it divides the training data into two subsets D_1 and D_2 and the predictive models trained on D_1 and D_2 improve the predictive performance in comparison to the model trained on the whole training data. To this end, we define useful contexts as follows:

Definition 1. (Useful Context). *Given a model M built based upon the whole training data D and M_1, M_2 are two models built based upon D_1 and D_2 corresponding to each contextual category of a context C respectively. The context C is useful if an only if: $E[T_1, M_1] \geq E[T_1, M]$ and $E[T_2, M_2] \geq E[T_2, M]$*

Essentially, this definition captures the usual operational situation in which no global optimum is sought, but there is a current system (captured by model M) that we seek to improve by taking into account context C.

4.3   Location as Context

Next, we will discuss what is the impact of geographical location as a contextual information. The geographical location of users is one of the prototypical aspects as a contextual information. In the literature, it was shown that the *user's location* is useful contextual information in many applications [5,31,36]. A context based on geographical location can have different levels of granularity like continent, country, city and so on.

In our experiments with StudyPortals [26], we consider a task of users' next action prediction. In order to accomplish this task we build contextual Markov models. We concentrate on a continent level of geographical location due to limitations from the data size side. We use users IP addresses as contextual features, then $\theta_s = IP$ is contextual vector associated with each user session s. We define six contextual categories: $C_{geo} = \{C_1 = Europe, C_2 = Africa, C_3 = North\ America, C_4 = South\ America, C_5 = Asia, C_6 = Oceania\}$. We have shown in [26] that for the case of StudyPortals the geographical location is no useful context.

Geographical location on a city level is considered as a context in TREC Contextual Suggestion Track [9,10,14]. The main goal of this task is to learn user's preferences out of provided examples of users' profiles where users rate different attractions. Afterwards, we need to return a ranked list of up to fifty ranked suggestions for each pair of user profile and context. The list of suggestions is ranked based on the user's preferences in the particular geographical location. As a source for contextual suggestions we used data from four social networks namely Facebook, Foursquare, Yelp, and Google Places, which are combined into one dataset. In order to achieve this goal, Kiseleva et al. [23] formulated the problem setup as a learning to rank problem where we directly optimize the required evaluation metrics, e.g., precision at rank 5 ($P@5$). We showed that our approach can be used in a preselection phase of contexts in the contextual suggestion task, but also that location is not equally useful for all web applications.

4.4   User Behavioral Aspects as Context

In addition to the relatively static location context, we will now look at dynamic context and discuss how to discover users behavioral aspects as contextual information. In case of StudyPortals[6] [25–27], users historical behavior is given as a log of web sessions corresponding to historical browsing activities of a user. In our case the users' actions are categorized by the type of the users' actions: searches, clicks on ads or homepage visits. Users' activities and their possible orderings within user web sessions is summarized as a user navigation graph. We want to understand if there are any groups of nodes in the navigation graph and then use this knowledge to characterize the users' behaviour in order to improve effectiveness of next users' action prediction. In order to achieve our goal we propose to use several machine learning techniques: First, we discover two types of user's behaviour on a site by grouping the user navigation graph: (1) expert users, who is experienced with website interface or searches extensively to find required information, and (2) novice user, who needs more time to learn about a website or is not interested much in content. Second, we discover changes in user intents while browsing a website. In order to achieve it we apply hierarchical clustering techniques with different optimisation functions: (1) directly maximizing the accuracy of next action prediction [25], and (2) directly minimizing the compression length [27] of decomposed web sessions. We described how the discovered contexts can utilized for the benefits of particular applications and use cases.

4.5   Changes in User Behavior over Time

In the final part we will look at changes in context over time, and discuss how to define and to detect changes in user satisfaction with retrieved search results. We look at indicators of a drop in user satisfaction due to SERPs trained on historical data becoming dis-aligned with a drift in query intent over time [22, 24].

When users struggle to find an answer for query Q they run a follow-up query Q' that is an expansion of Q. Query reformulation is the act of submitting a next query Q' to modify a previous $SERP$ for a query Q in the hope of retrieving better results [18]. Such a query reformulation is a strong indication

---

[6] See: http://www.studyportals.eu/.

of user dissatisfaction [3]. We call this the *reformulation signal*. Our hypothesis is that a decrease in user satisfaction with $\langle Q, SERP \rangle$ correlates nicely with the reformulation signal. In other words, the probability of reformulating Q will grow dramatically.

We propose an unsupervised approach, called *Drift Detection in user SATisfaction (DDSAT)*, for detecting drifts in user satisfaction for pairs $\langle Q, SERP \rangle$ by applying a concept drift technique [38] leveraging reformulation signal. Concept drift primarily refers to an online supervised learning scenario when the relation between the input data and the target variable changes over time [16]. Furthermore, the reformulation signal is considered to be less noisy and if reformulations are fresh and done only by users' initiative then we can say that a reformulation signal is not biased by information coming from the search engine.

We conduct a large-scale evaluation using search log data from Microsoft Bing[7] [22] and Yandex[8] [24] where we extend our framework by taking into account more signs of user frustration (lack of search satisfaction) such as: a rate of search abandonment, a dramatical change in query volume, a lowering in average click positions. Our experiments show that the algorithm *DDSAT* works with a high accuracy. Moreover, our framework outputs the list of drift terms and the list of $URLs$, which can be used for the future re-ranking of $SERP$. The algorithm of the drift detection in user satisfaction can be incorporated in many search-related applications where freshness is required, e.g. in recency ranking, query auto-completion.

In addition, we conducted conceptual analysis to clarify the meaning of core concepts and their relations and dependencies. And as a conceptual model, we worked with an idealized model that abstracts away from other factors outside the scope of our interest.

In the next section, we discuss the setup and results of the hackathon.

5   The Hackathon

The hackathon consisted of a miniature version of the TREC Contextual Suggestion Track. The stated goal of the hackathon was to provide recommendations to the organizers of this tutorial regarding sites to see and things to do in Saint Petersburg during their visit. The hackathon was initiated the evening of Tuesday, August 25, with pizza and beer provided by the organizers as inspiration. Teams reported out two days later, with presentations on the evening of Thursday, August 27. The hackathon involved ten teams, with a total of 30 participants.

No time was allocated for working on the task during the days in between—only evenings were available, limiting the amount of work that could be done. Nonetheless, as described below, a number of teams put considerable effort into the task, coming up with many highly creative and interesting solutions. After the report-out, the organizers awarded prizes for Best System, Best Presentation,

---

[7] See: http://www.bing.com/.

[8] See: http://yandex.ru/.

Table 2. Context mapping from IDs to Cities and States

Id	City	State
151	New York City	NY
152	Chicago	IL
⋮		
421	Walla Walla	WA
422	Lewiston	ID
423	Saint Petersburg	Russia

and Most Original Approach. Slides from some of the student presentations are available online.[9]

5.1 Data Resources Available

The hackathon used a variant of the TREC 2015 Contextual Suggestion Track's Batch Task,[10] tailored to the lecturers visiting St. Petersburg. The data is available from http://plg.uwaterloo.ca/~claclark/russir2015/.

Data. First, there is the core data material: the requests (input) and sample responses (the results your system should generate). In short: you get a new context (city) with candidate venues to rank, plus detail about the person asking, including what she/he likes in other cities. The contexts are a simple csv file with `id`, `city`, and `state` fields, as shown in Table 2. The contexts are based on the TREC contextual suggestion track (US cities) extended with St. Petersburg. The lecturers apologized for the format which makes the inappropriate suggestion that Russia is a US state. The requests consists of the profile and context, as well as the candidates to rank as shown in Table 3. It details the request (901), the context (location), and details about the person requesting (person) and the trip, as well as the candidates to rank. The profile contains a large set of preferences n another context or city. The responses consists of the group and run details, as well as a ranked list of suggestions (derived from the candidates) as shown in Table 4.

Evaluation. Second, there is an evaluation package for the US cities, to evaluate or train systems. This consists of the judgments by the person for each of the candidates (ratings and tags/endorsements), a script to transform the response file into the TREC format, and the TREC script to calculate standard IR measures.

---

[9] http://plg.uwaterloo.ca/~claclark/russir2015/Students.
[10] See: https://sites.google.com/site/treccontext/.

Table 3. Request: Profile and candidates to rank

JSON request

```
{ "body" : {
     "group" : "Friends",
     "duration" : "Longer",
     "season" : "Autumn"
     "trip_type" : "Holiday",
     "person" : {
        "preferences" : [
           {"documentId" : "TRECCS-00247656-160",
              "tags" : [
                 "Bar-hopping",
                 "Clubbing"
              ],
              "rating" : "4"
           },
           {"documentId" : "TRECCS-00211603-161",
              "tags" : [
                 "Fast Food",
                 "Restaurants"
              ],
              "rating" : "0"
           },
           ...
        ],
        "id" : 1234568,
        "age" : "47",
        "gender" : "male"},
     "location" : {
        "id" : 423,
        "lat" : 59.95, "lng" : 30.3,
        "name" : "Saint Petersburg"},
  },
  "id" : 901,
  "candidates" : [
     "TRECCS-00000001-423",
  ...
     "TRECCS-00000102-423"]}
```

Additional Data. Third, there is additional data that can be used. There is the crawled page of the venues (all URLs of venues in the Batch task, as well as those in St. Petersburg). There is also the data used at TREC about a much larger set of pages, as detailed on the TREC pages. And there are categories and ratings of the US venues, obtained from a commercial service.

Table 4. Response: group and run details plus suggestions.

JSON response

```
{
    "groupid" : "demo",
    "runid" : "demoA",
    "id" : 901,
    "body" : {
        "suggestions" : [
                "TRECCS-00000099-423",
                "TRECCS-00000006-423",
                ...
                "TRECCS-00000079-423"         ]
    }
}
```

5.2   Student Presentations

To provide a sense of the breadth and variety of approach, we provide a short overview of the efforts from several groups, who were kind enough to provide slides and other material after the hackathon.

- Team **MAD IT** (Maria Zagulova, Andrey Poletaev, Dmitry Zhelonkin, Ivan Grechikhin, and Tania Nikulina) clustered people on the basis of demographics (age, gender, etc.) and personalized each cluster using tag activity. For one organizer, suggestions included Le Tour de Vin wine bar, Saint Petersburg 300 Year Park, and well as various cafes. For the other two organizers, suggestions included the Grand Market Russia, the Faberge Museum, and the Mikhailovsky Theatre, as well as a tattoo parlor. These suggestions were well received by the organizers, with the exception perhaps of the tattoo parlor. The team has made source code, as well as more information about their work available for interested readers[11].
- Team **No Name** (Michael Nokel) took a colaborative filtering approach, by throwing out all data other than the attraction ratings and applying singular value decomposition (SVD). Gradient boosting regression was then applied to a combination of user features and SVD features, showing improvements over SVD along on the training data. Unfortunately, no recommendations for the organizers were made.
- Team **Rambler & Co** (Maria Zagulova, Andrey Poletaev, Dmitry Zhelonkin, Ivan Grechikhin, and Tania Nikulina) used vowpalwabbit rank approximations to predict user ratings. Features included gender, age, season, etc., as well as LDA topics of Tripadvisor and Foursquare titles trained on translated titles. For candidate attractions in Saint Petersburg, the team manually assigned tags. Suggested attractions incuded the El Copitas Cocktail Bar, the Wine Bar Bratya Tonet, and the Co-op Garage Bar.

---

[11] bitbucket.org/poletaev/russir-2015/src.

– Team **sleep_deprived** (Sagun Pai and Sheikh Muhammad Sarwar) applied collaborative filtering methods, approaching the cold start problem through an tag expansion approach, for example, automatically expanding the tag "food" to include "seafood". An expanded vector of tags was created for each user, and these expanded vectors were used to select recommendations within Saint Petersburg. Suggestions included some of the same attractions recommended by other groups, (e.g., the Mikhailovsky Theater), as well as various bars (8th Line Pub), restaurants (Wave Burgers), and palaces (Catherine Palace).

5.3 The Outcome

After student presentations were complete, prizes were awarded as follows:

– the *Best System Award* went to MAD IT (Maria Zagulova, Andrey Poletaev, Dmitry Zhelonkin, Ivan Grechikhin, and Tania Nikulina);
– the *Best Presentation Award* went to SalsaRoulette (Navid Rekabsaz, Larisa Adamyan, Ioanna Miliou, and Aldo Lipani); and
– the *Most Original Approach Award* went to sleep_deprived (Sagun Pai, Sheikh Muhammad Sarwar).

Congratulations! And thank you to all groups for making the experience so enjoyable. The organizers visited several of the recommended places over the weekend, and can confirm that the suggestions were indeed highly relevant.

In the next section, we concludes the paper with some discussion on the outcome of the lectures and hackathon.

6   Conclusion

This paper described the course on contextual search and exploration, given as part of the ninth Russian Summer School in Information Retrieval (RuSSIR 2015).[12] Our goal was to enable students to understand the challenges and opportunities of contextualized search over entities, and learn effective approaches for the concrete application to venue recommendation domain, as well as obtain hands-on experience with developing and evaluating personalized search and recommendation approaches.

The course consisted of both lectures and hands-on "hackathon" sessions with data derived from the TREC task. First, we gave an overview of approaches to contextual search and exploration, both in terms of the general problem of complex task support, and specifically focusing on a venue recommendation task. Second, we detailed how to set up an experiment to evaluate contextual suggestion based on the TREC track. Third, we detailed a general approach to using contextual information in modeling search and interaction behavior. Fourth, we discussed the setup and results of the hackathon, in which the students were asked to make recommendations to the lecturers on what to do in St. Petersburg after the course.

---

[12] See: http://romip.ru/russir2015/.

Acknowledgments. We are grateful to RuSSIR to cover the travel expenses of the first author. We are thankful to the 30 students that actively participated in the hackathon—we were deeply impressed by the amount of work and creative ideas that were tried within 48 hours!

References

1. Abowd, G.D., Dey, A.K.: Towards a better understanding of context and context-awareness. In: Gellersen, H.-W. (ed.) HUC 1999. LNCS, vol. 1707, pp. 304–307. Springer, Heidelberg (1999)
2. Adomavicius, G., Tuzhilin, A.: Context-aware recommender systems. In: CARS (2010)
3. Ageev, M., Guo, Q., Lagun, D., Agichtein, E.: Find it if you can: a game for modeling different types of web search success using interaction data. In: SIGIR (2011)
4. Allan, J., Croft, B., Moffat, A., Sanderson, M.: Frontiers, challenges, and opportunities for information retrieval: report from SWIRL 2012 the second strategic workshop on information retrieval in Lorne. SIGIR Forum **46**(1), 2–32 (2012)
5. Alves, A.O., Pereira, F.C.: Making sense of location context. In: Proceedings of the 1st International Workshop on Context Discovery and Data Mining (ContextDD 2012), vol. 4, p. 7. ACM, New York (2012). http://dx.doi.org/10.1145/2346604.2346609
6. Bolchini, C., Curino, C.A., Quintarelli, E., Schreiber, F.A., Tanca, L.: A data-oriented survey of context models. In: SIGMOD (2007)
7. Brown, P., Bovey, J., Chen, X.: Context-aware applications: from the laboratory to the marketplace. IEEE Pers. Commun. **4**, 58–64 (1997)
8. Cao, H., Hu, D.H., Shen, D., Jiang, D., Sun, J.-T., Chen, E., Yang, Q.: Context-aware query classification. In: SIGIR (2009)
9. Dean-Hall, A., Clarke, C.L., Kamps, J., Thomas, P., Voorhees, E.: Overview of the TREC 2014 contextual suggestion track. In: 23rd Text REtrieval Conference, Gaithersburg, Maryland (2015)
10. Dean-Hall, A., Clarke, C.L., Simone, N., Kamps, J., Thomas, P., Voorhees, E.: Overview of the TREC 2013 contextual suggestion track. In: 22nd Text REtrieval Conference, Gaithersburg, Maryland (2014)
11. Dean-Hall, A., Clarke, C.L.A.: Assessing contextual suggestion. In: 6th International Workshop on Evaluating Information Access, Tokyo, December 2014
12. Dean-Hall, A., Clarke, C.L.A.: The power of contextual suggestion. In: 37th European Conference on Information Retrieval, pp. 352–357, Vienna, March 2015
13. Dean-Hall, A., Clarke, C.L.A., Kamps, J., Thomas, P.: Evaluating contextual suggestion. In: 5th International Workshop on Evaluating Information Access, Tokyo, June 2013
14. Dean-Hall, A., Clarke, C.L.A., Kamps, J., Thomas, P., Voorhees, E.: Overview of the TREC 2012 contextual suggestion track. In: 21st Text REtrieval Conference, Gaithersburg, Maryland (2013)
15. Dey, A., Abowd, G., Salber, D.: A conceptual framework and a toolkit for supporting the rapid prototyping of contextaware applications. Hum. Comput. Interact. **2**, 97–166 (2001)
16. Gama, J., Žliobaitė, I., Bifet, A., Pechenizkiy, M., Bouchachia, A.: A survey on concept drift adaptation. ACM Comput. Surv. **46**, 4:1–4:37, Article 44 (2014). http://dx.doi.org/10.1145/2523813

17. Hashemi, S.H., Clarke, C.L., Dean-Hall, A., Kamps, J., Kiseleva, J.: On the reusability of open test collections. In: 38th International ACM SIGIR Conference on Research and Development in Information Retrieval, Santiago, Chile, pp. 827–830, August 2015

18. Hassan, A., Shi, X., Craswell, N., Ramsey, B.: Beyond clicks: query reformulation as a predictor of search satisfaction. In: CIKM, pp. 2019–2028 (2013)

19. Hull, R., Neaves, P., Bedford-Roberts, J.: Toward situated computing. In: ISWC, pp. 146–153 (1997)

20. Kiseleva, J.: Context mining and integration into predictive web analytics. In: WWW (Companion Volume), pp. 383–388 (2013)

21. Kiseleva, J.: Using contextual information to understand searching and browsing behavior. In: Submission of SIGIR (Doctoral Consorcium) (2015)

22. Kiseleva, J., Crestan, E., Brigo, R., Dittel, R.: Modelling and detecting changes in user satisfaction. In: Proceeding of CIKM, pp. 1449–1458 (2014)

23. Kiseleva, J., García, A.M., Luo, Y., Kamps, J., Pechenizkiy, M., Bra, P.D.: Applying learning to rank techniques to contextual suggestions. In: Proceeding of Text REtrieval Conference (TREC) (2014)

24. Kiseleva, J., Kamps, J., Nikulin, V., Makarov, N.: Behavioral dynamics from the SERP's perspective: what are failed SERPS and how to fix them? In: CIKM, pp. 1561–1570 (2015)

25. Kiseleva, J., Lam, H.T., Pechenizkiy, M., Calders, T.: Discovering temporal hidden contexts in web sessions for user trail prediction. In: Proceedings of WWW (Companion Volume), pp. 1067–1074. ACM (2013)

26. Kiseleva, J., Lam, H.T., Pechenizkiy, M., Calders, T.: Predicting current user intent with contextual markov models. In: ICDM Workshops (2013)

27. Lam, H.T., Kiseleva, J., Pechenizkiy, M., Calders, T.: Decomposing a sequence into independent subsequences using compression algorithms. In: Proceedings of the ACM SIGKDD Workshop on Interactive Data Exploration and Analytic, pp. 67–75 (2014)

28. Palmisano, C., Tuzhilin, A., Gorgoglione, M.: Using context to improve predictive modeling of customers in personalization applications. IEEE Trans. Knowl. Data Eng. (TKDE) **20**(11), 1535–1549 (2008)

29. Prahalad, C.: Beyond CRM: predicts customer context is the next big thing. In: AMA MwWorld (2004)

30. Rendle, S., Gantner, Z., Freudenthaler, C., Schmidt-Thieme, L.: Fast context-aware recommendations with factorization machines. In: Proceeding of the 34th International ACM SIGIR Conference on Research and Development in Information Retrieval, SIGIR 2011, pp. 635–644 (2011)

31. Schmidt, A., Beigl, M., Gellersen, H.-W.: There is more to context than location. Comput. Graph. **23**(6), 893–901 (1999)

32. Smucker, M.D., Clarke, C.L.: Time-based calibration of effectiveness measures. In: 35th International ACM SIGIR Conference on Research and Development in Information Retrieval, Portland, Oregon, pp. 95–104 (2012)

33. Stern, D., Herbrich, R., Graepel, T.: Matchbox: large scale online bayesian recommendations. In: WWW, pp. 111–120 (2009)

34. Turney, P.: Exploiting context when learning to classify. CoRR (2002)

35. Turney, P.: The management of context-sensitive features: a review of strategies. CoRR (2002)

36. Want, R., Hopper, A., Falcão, V., Gibbons, J.: The active badge location system. ACM Trans. Inf. Syst. (TOIS) **10**(1), 91–202 (1992)

37. Xiang, B., Jiang, D., Pei, J., Sun, X., Chen, E., Li, H.: Context-aware ranking in web search. In: SIGIR (2010)
38. Zliobaite, I.: Learning under concept drift: an overview. CoRR abs/1010.4784 (2010)
39. Žliobaitė, I.: Identifying hidden contexts in classification. In: Huang, J.Z., Cao, L., Srivastava, J. (eds.) PAKDD 2011, Part I. LNCS, vol. 6634, pp. 277–288. Springer, Heidelberg (2011)
40. Zliobaite, I., Bakker, J., Pechenizkiy, M.: Beating the baseline prediction in food sales: how intelligent an intelligent predictor is? Expert Syst. Appl. (ESWA) **39**(1), 806–815 (2012)

A Tutorial on Leveraging Knowledge Graphs for Web Search

Gianluca Demartini[✉]

University of Sheffield, Sheffield, UK
g.demartini@sheffield.ac.uk

Abstract. Knowledge Graphs are large repositories of structured information about entities like persons, locations, and organizations and their relations. Modern Web search engines leverage such background Knowledge Graphs to create rich search engine result pages for entity-centric search queries.

In this document we provide an introduction to Knowledge Graphs and their application to search-related problems. We present techniques to search for entities instead of documents as answer to a search query. Finally we present human computation techniques to build hybrid human-machine systems to solve entity-oriented search tasks making use of Knowledge Graphs.

1 Introduction

Web search engines have evolved beyond just presenting the classic ten blue links in the search engine result page (SERP) as an answer to a keyword query. Modern Web search engines include in the SERP results from verticals such as news, images, videos, etc. More than that, Web search engines present rich search result pages when users query for information about specific entities like actors or movies. Depending on the type of entity users ask for, the structure of the information presented differs. Search engine result pages may include news articles, pictures, factual statements, and related entities. This is due to the fact that users of Web search engines look for specific entities on-line. Indeed, about 50 % of the query workload a commercial search engine receives is related to specific entities [20].

Information presented in such rich search engine result pages is taken at query time from a background Knowledge Graph by matching the user query against all entities stored in the Knowledge Graph. Once the relevant entity has been identified, information is retrieved from the Knowledge Graph and presented to the user.

In this document we provide an overview of the different steps involved in creating such entity-centric user experiences in Web search engines.

In Sect. 2 we present the fundamental definitions of syntax and query language for Knowledge Graphs. We describe the Relational Description Framework (RDF) as well as the SPARQL query language to access data stored in Knowledge Graphs.

© Springer International Publishing Switzerland 2016
P. Braslavski et al. (Eds.): RuSSIR 2015, CCIS 573, pp. 24–37, 2016.
DOI: 10.1007/978-3-319-41718-9_2

In Sect. 3 we describe the standard approaches to extract and uniquely identify entities such as persons, locations, and organizations in textual content such as, for example, news article or general Web pages.

In Sect. 4 we introduce systems that make use of entities to provide search functionalities to Web users. We describe indexing and ranking approaches related to a number of entity-oriented search tasks.

In Sect. 5 we discuss hybrid human-machine systems that make use of crowdsourcing to deal with Knowledge Graph related problems such as entity linking, data integration, and entity search.

Finally, Sect. 6 concludes this document highlighting open direction for future research.

2 Introduction to Knowledge Graphs

Rich SERPs containing entities are possible thanks to what goes under the name of *Web of Data* and Knowledge Graphs. In the case of Google, the Knowledge Graph project launched in May 2012 following the acquisition of Metaweb in July 2010 with its main product Freebase: a freely editable Knowledge Graph available both in human-readable and machine-readable formats.

Similarly, other Knowledge Graphs have been created by the academic community under the label of *Linked Open Data*[1] (LOD). Most popular datasets in LOD include DBpedia[2], Freebase[3], and YAGO[4]. DBpedia is a collection of RDF triples automatically extracted from Wikipedia article leveraging the semi-structured information present in it (e.g., the Wikipedia info-boxes) and manually defined mapping rules that allow to extract structured data out of Wikipedia articles.

In the commercial domain, Knowledge Graphs have been developed and used to power end-user applications such as Web search. Examples include the Google Knowledge Vault [15], the Facebook Entity Graph[5], and Microsoft Satori which is used, among other applications, to power rich search engine result pages.

2.1 Information Extraction and Knowledge Acquisition

The process in which information is extracted from unstructured text to create Knowledge Graphs is called *Information Extraction and Knowledge Acquisition*. Starting from textual data, information extraction techniques will identify structured components like, for example, entities and factual statements. Then, the process of knowledge acquisition connects factual statements generating a Knowledge Graph by applying techniques for fact consistency and data integration.

---

[1] http://linkeddata.org.
[2] http://dbpedia.org.
[3] https://www.freebase.com/.
[4] https://www.mpi-inf.mpg.de/departments/databases-and-information-systems/
research/yago-naga/yago/.
[5] https://www.facebook.com/notes/facebook-engineering/
under-the-hood-the-entities-graph/10151490531588920.

Early information extraction solutions were based on manually defined patterns matched against textual documents in order to identify expected occurrences of statements. Early solutions used to construct Knowledge Graphs have been based on manual effort like, for example, Cyc [22] which has collected 250 thousand entities over 20 years. The objective of current work is the combination of information extraction techniques to automatize and scale-up the knowledge acquisition process.

In summary, we can group approaches to construct Knowledge Graphs into manually-supported construction (e.g., Freebase and Wikidata) where human intervention is used to complement automatically extracted data, and automatic methods. Among automatic methods we can find Knowledge Graphs extracted from semi-structured data like DBpedia and YAGO which leverage existing structure in Wikipedia, Knowledge Graphs extracted from text, and Knowledge Graphs resulting from the combination of text extraction and database integration like the Google Knowledge Vault [15].

When creating Knowledge Graphs a number of challenges have to be tackled. These include the choice about which entities to include and which not to include. As an example, Knowledge Graphs based on Wikipedia follow the *notability criteria*[6] which states that only popular entities should be included. Thus, these Knowledge Graphs trade-off high data accuracy for low coverage. Another challenge is about keeping information in the Knowledge Graph up-to-date. For this, two alternatives are usually adopted: either outdated information is modified or factual statements are annotated with a time validity interval (e.g., the fact 'George W Bush is president of the USA' has a time validity between 2001 and 2009).

2.2   RDF Data Format

The W3C standard to define structured information on the Web is RDF[7]. Data in LOD is typically available in RDF format to download or to access over SPARQL endpoints as answer to a query.

RDF, standing for Resource Description Framework, encodes data as triples in the form of *subject*, *predicate*, and *object* which naturally form distributed graphs. This data format can be used to describe semi-structured information and factual statements (e.g., subject:Tom_Cruise predicate:Starring_In object:Top_Gun). Entities appearing as subject or object of RDF statements are represented by Unique Resource Identifiers (URIs) in order to relate all statements about the same entities. Additionally to that, the object of an RDF statement can assume a numerical or literal value instead of an entity (e.g., subject:Tom_Cruise predicate:Born_In object:1962).

Data in RDF format can be encoded in different ways. Traditionally, XML serializations were used to store and exchange RDF data. Recently, the use of

---

[6] https://en.wikipedia.org/wiki/Wikipedia:Notability.
[7] http://www.w3.org/RDF/.

JSON-LD[8], a variant of the popular JSON format, has become popular to encode such factual statements about entities and their properties.

In order to define data in RDF we need to follow a schema, that is, a common vocabulary used to define attributes of entities (e.g., 'starring_in'). To this aim, RDF Schemas (RDFS) are defined to be re-used. RDFS define properties and classes to be used for a certain domain and have been defined by the corresponding community. Popular examples, of RDFS include *foaf*[9] to define relations in social networks, *Dublin Core*[10] to describe publications, and *Good Relations*[11] used to describe products. RDFS constructs include *classes* (i.e., types of entities) and *properties* of entities. It is also possible to create hierarchies by using *SubClassOf* and *SubPropertyOf* constructs. Finally, it is possible to create constraints on the class of subject and object for a certain predicate by defining the predicate *Domain* and *Range* respectively.

Once data is represented in RDF format and stored in an *RDF store*, structured queries can be run against it to retrieve specific subsets of the data. Similar to SQL for Databases, RDF data can be queried using SPARQL[12]: A declarative query language for RDF and RDFS. SPARQL leverages the fact that RDF is represented as triples and defines *triple patterns* to be matched against the data. Like for SQL, a number of constructs exists for SPARQL including 'order by', 'distinct', 'limit', etc.

3   Named Entity Recognition and Linking to Knowledge Graphs

In this section we describe a classic Natural Language Processing (NLP) pipeline consisting of Named Entity Recognition (NER) followed by Entity Linking (i.e., entity disambiguation). Next, we describe one additional step on top of such NLP pipeline which consists of the selection of entity types from the Knowledge Graph to be displayed to users reading a document where a certain entity appears.

The first step to process unstructured documents to create a connection to an existing Knowledge Graph is NER, also known as Entity Extraction. The step of NER consists of identifying entity mentions in textual documents. Thus, the result of NER will be the indication of which n-gram in the text represents an entity (e.g., a person, a location, or an organization). Next, the step of entity resolution or entity linking consists of uniquely identifying the extracted entity by creating a link to an entity described in an external Knowledge Graph. In this way, we 'resolve' the extracted n-gram to one specific entity and we are able to uniquely identify it. This requires also the step of entity classification, that is, distinguishing among different types of entity.

---

[8] http://json-ld.org/.
[9] http://www.foaf-project.org/.
[10] http://dublincore.org/.
[11] http://www.heppnetz.de/projects/goodrelations/.
[12] http://www.w3.org/TR/rdf-sparql-query/.

3.1 Named Entity Recognition

In order, the steps of classic NLP pipelines are: tokenization, sentence splitting, part-of-speech (POS) tagging, NER, entity-linking, co-reference resolution, and relation extraction. NER approaches make use of tokenization and POS tags and can be classified in the following categories: dictionary-based, pattern-based, and supervised learning models. While the simpler approaches based on dictionaries and patterns work well in certain domains like, for example, for geographical entities where it is easy to obtain a comprehensive list of existing entities, most advanced techniques make use of supervised machine learning models. In detail, NER can be formulated as a classification task where the goal is to assign to each token the tag B (i.e., beginning of an entity), I (i.e., continuation of an entity), or O (i.e., word outside an entity). Next, a classification problem can be defined to attach a type to an entity (e.g., person, organization, etc.). Common machine learning models used for these problems are Decision Trees and Support Vector Machines. Finally, the most popular NER approaches use probabilistic sequence models where Hidden Markov Models and Conditional Random Fields are the most widely used. In this type of approaches each token in a sequence is assigned to a label which is dependent on the labels of tokens in its proximity. Popular features used by supervised models used for NER include gazetteers, orthographic features (e.g., capital letters), word types, POS tags, context, and trigger words (e.g., 'Mr', 'Miss'). Software libraries that implement state of the art NER approaches have been developed. Popular libraries include the Stanford Named Entity Recognizer[13] for Java and NLTK[14] for Python.

3.2 Entity Linking

Once named entities have been identified, the next step is to use the entity mention to gather candidate entities from the Knowledge Graph. Then, by producing a ranked list of candidate matching entities, we can select the best match for the entity mentioned in the document. More than text similarity measures, it is possible to leverage the context in which entities appear to obtain better disambiguation. The intuition is that entities that co-occur in the same textual context tend to be more related each other. To leverage this intuition we need to express such relatedness in a numerical way by, for example, measuring statistical co-occurence, similarity of entity descriptions in the Knowledge Graph, or measuring the distance of the entities following relations in the Knowledge Graph. As an example of relatedness measured by co-occurence, we can imagine a document mentioning both the entity 'FC Barcelona' and the entity 'Bayern'. Based on textual match, the entity 'Bayern' could match to either the entity 'Bavaria' (i.e., the region in Germany) or the entity 'Bayern München' (i.e., the football team) in the Knowledge Graph. As from other documents in the collection we can observe that the entity 'FC Barcelona' (i.e., a football team) often co-occurs with the entity 'Bayern München', we select this option as the most likely match for the mention extracted from the document.

---

[13] http://nlp.stanford.edu/software/CRF-NER.shtml.
[14] http://www.nltk.org.

More than just general text documents on the Web like, for example, news articles, entities may appear in other textual content. Recent research focus has been put on identifying entities appearing in structured tables on the Web [16], micro-blogs [19], and Web search queries [20,24] where adapted approaches have been designed.

Recent work has shown that about 70 % of Web search queries contain a named entity, that about 50 % of queries have an entity focus, and that 10 % of queries are looking for a certain entity type [20]. In these cases, the query intent is typically expressed by words additionally to the mentioned entity used to disambiguate it, for example, by expressing its type (e.g., 'tom cruise actor'). NER approaches developed to work best on search queries look at certain keyword matches [3] and at looking up entity names in the Knowledge Graph [5]. As a demonstration of the interest of the Information Retrieval (IR) community to NER and Entity Linking, at SIGIR 2014 the Entity Recognition and Disambiguation Challenge was run to disambiguate entities appearing in Web search queries and Web pages to entities in the Freebase Knowledge Graph.

3.3 Ranking Entity Types

Once entities appearing in documents have been extracted and disambiguated by creating links to entities in the Knowledge Graph, we can start to create end-user applications that make use of the information available in the Knowledge Graph. As an example application, we now describe a system to select the most relevant type of an entity mentioned in a document to be showed to users reading the document as an explanation of the entity they are reading about [29].

Once an entity mention in a document is disambiguated to one entity appearing in a Knowledge Graph, a number of factual statements about that entity are available. One piece of information is the *type* of the entity. For example, for the entity 'Tom Cruise', DBpedia shows more than 40 different types attached to it including 'Person', 'Artist', 'Actor', and 'Producer'. The task is to select one of these types as the most informative for a user to see when reading a document mentioning Tom Cruise. While it is clear that the type Person may, in most of the cases, be not so informative as it is too general, the selection between Actor and Producer may depend on the textual context where the entity appears.

The NLP pipeline developed for this [29] looks similar to the classic NLP pipeline up to the task of entity linking. After this, a supervised machine learning model is used to produce a ranked list of entity types obtained from the Knowledge Graph. Among the diverse set of features used to rank entity types, the most effective ones are based on the type hierarchy (i.e., information about the fact that Actor is a sub-type of Person) and on the types of other entities co-occuring with the entity for which we are ranking types.

When applying such NLP pipelines more than just accurate results we would like the approaches to work at Web scale. In [29] the proposed pipeline (which is available at https://github.com/MEM0R1ES/TRank) has been deployed over an Hadoop cluster using a Map/Reduce implementation. By building inverted indices on top of the Knowledge Graph to store entity description used for entity linking and to store the entity type hierarchy used to select entity types, it is possible to distribute such data structures on each node of the cluster and run the pipeline in parallel. The specific implementation could run the pipeline at a rate of 100 documents per node per second. Interestingly, the split of the overall execution time among the different steps in the pipeline indicates that the most time consuming step is NER (with 36 %) while the least time consuming step is type ranking with 6 % of the overall time spent to process a document.

4   Searching for Entities

As mentioned in Sect. 3 searching for entities is becoming more and more popular over time for Web search engine users. In this section we describe different approaches for entity search. We start from the historically first entity type for which search systems have been developed, that is, persons, by describing how expert finding systems work. Then, we discuss general entity ranking on top of a Knowledge Graph. Next, we talk about the popular ad-hoc object retrieval task (i.e., given a keyword query describing an entity, retrieve the matching entity from a Knowledge Graph).

4.1   Expert Finding

Expert finding is defined as the search task of returning a list of candidate experts ranked by expertise on the topic described by the user query [1]. Expert finding systems are most valuable in a large enterprise setting where skills and competencies are spread across departments. In a scenario where executives needs to create teams to work on new projects or simply to find the right person to solve a problem, expert finding systems can produce a ranked list of company employees on a certain topic described by the user query. Such ranked list is generated based on the digital content available in the enterprise after being appropriately indexed.

We distinguish two basic expert finding approaches: *document-centric* and *candidate-centric* approaches.

In document-centric approaches an inverted index is created on top of the documents available within the enterprise. Then, given a user query a ranked list of documents matching the user query is generated using classic IR ranking models. Finally, candidate experts are extracted from top ranked documents to generate a list of experts to be returned as answer to the query.

In candidate-centric approaches, names of candidates are extracted from all documents first. Then, candidate profiles are created by aggregating information from all documents where the candidate name appears. Finally, an index is created over candidate profiles and, given a user query, a ranked list of candidate profiles is generated as answer.

Other models for expert finding exist. One example include voting models where, by means of data fusion techniques, ranked documents are seen as votes to candidate expertise. Votes are than aggregated by taking into account different features like, for example, document rank, score, or the number of different documents voting for the same candidate [21]. Another example of expert finding are user-oriented models where additional real-world constrains are taken into account. Possible dimensions are the user previous knowledge on the topic she is looking experts for (i.e., retrieved experts should have wider knowledge than the user) and contact time (i.e., taking into account geographical location of the user and the candidate experts or their distance in the organizational hierarchy) [27].

4.2 Entity Ranking

Generalizing expert finding techniques to any entity type, we talk about entity ranking systems. Thus, we aim at building systems which can, at the same time, answer queries about, for example, 'Impressionist art museums in Holland' as well as about 'German car manufacturers'. Because of the availability of multiple entity types, Wikipedia has been used as main background collection for entity ranking systems. Over time, effective approaches have strongly leveraged the category structure of Wikipedia to identify relevant entity types, looked at query extension techniques by means of synonyms and other related words, and built on top of the link structure connecting Wikipedia pages to identify alternative entity labels in the anchor text of hyperlinks [10].

More than searching for entities over a static collections like Wikipedia, entity ranking approaches are useful when ranking entities in a collection evolving over time. The most relevant example is that of news articles that are published over time about the same story. In the case of news events that evolve over time entities may appear only later in the story or be present at the beginning but than become not relevant anymore. Thus, in such cases we want entity ranking systems that leverage past information to generated better rankings [11].

4.3 Ad-hoc Object Retrieval

A different entity search task is Ad-hoc Object Retrieval (AOR) [24] which is highly related to the NLP task of Entity Linking. In this task, a user keyword query describing one specific entity is used to generated a ranked list of entity identifiers retrieved from a Knowledge Graph. The goal is to identify which entity the user is looking for in order to display him with some structured information about the entity as retrieved from the background Knowledge Graph. This is typically done as a way to support end users who cannot express their information need using SPARQL and by giving up query expressivity to scale over large amounts of data. Moreover, users are not aware of the RDFS schema used to store and describe the data in the Knowledge Graph and thus there is the need to match any user keyword query against the available data.

In order to efficiently answer such queries without need to scan the entire collection, we must construct an index over the data stored in the Knowledge Graph. Given the semi-structured nature of Knowledge Graph data (i.e., it contains the structure relating entities as well as textual descriptions of entities), we can leverage IR indexing techniques for unstructured content still maintaining the possibility to search over the structure present in the data. Two alternative techniques have been proposed [4]: Horizontal index structure and Vertical index structure. Horizontal indexing requires two indices: one for terms and one for properties in the Knowledge Graph. Thus, for each term we store on the same index position the properties where it appears. In this way we need to store just two indices which grow horizontally as more data is added to the index. The dictionary size is the number of unique terms plus the number of properties in the Knowledge Graph. Vertical indexing requires to store one index per property. Each index will store all the terms appearing for that property. Thus, this type of index grows vertically in the number of properties. While the dictionary in this case is smaller (i.e., the number of unique terms), a dataset with many distinct properties could become inefficient as many merging operations across indices may need to take place to answer a query. To answer a query over these types of indexing techniques, functions that rank over multiple indices can be used (e.g., BM25F [25]).

An alternative indexing solution is the use of IR inverted indices in combination with graph-based indices over which SPARQL queries can be run [30]. In this system, given a query, a ranked list of entities is generated first using standard approaches. Then, as a refinement step, top-k results are looked up in the graph in order to leverage the graph structure to re-rank results and obtain a more effective AOR system. While graph traversal operations generate some overhead in terms of computational complexity, the efficiency of the system remains acceptable to answer user queries in real-time. Recent work on AOR has looked at approaches that account for term dependencies in the case of multi-field entity descriptions like in the case of the AOR task [31].

5   Crowdsourcing for Knowledge Graphs

In this section we present the concept of *crowdsourcing* and explain how it can be used to improve the quality of systems that make use of Knowledge Graphs. We will look at crowd-based entity linking system as well as at the use of crowdsourcing to understand complex Web search queries.

5.1   Crowdsourcing

Crowdsourcing is defined as the use of human intelligence at scale to solve problems that are simple for humans to complete but still difficult for machine-based algorithms (e.g., image labelling, text summarization, translation, etc.). When applied to Knowledge Graphs, crowdsourcing makes use of micro-tasks that take from few seconds to a couple of minutes for an individual to complete. The execution of tasks happens in so called *crowdsourcing platforms* like, for example, Amazon MTurk[15] where the *workers* (i.e., the crowd) and the *requesters* (i.e., those who publish tasks to be completed) meet. Workers complete Human Intelligence Tasks (HITs) which are grouped in batches of similar tasks in exchange of a small monetary reward. The access to the platform from the requester point of view can typically be done by means of a website or programmatically over APIs.

---

[15] http://mturk.com.

At the moment, the most popular crowdsourcing platform is Amazon MTurk which was launched in 2005. This platform works as a marketplace where requesters publish batches of HITs to be completed and workers are free to pick among the available tasks those they wish to complete in order to gain the reward assigned by requesters. On top of the assigned reward the platform takes a transaction fee proportional to the reward and to the number of tasks published. In [13] an analysis of the evolution of this crowdsourcing platform over time is presented.

Crowdsourcing has been applied in a variety of areas in computer science. For example, the IR community has leveraged crowdsourcing as a means to obtain relevance judgements at scale. This is possible as crowd workers are cheaper and faster than the traditional assessors that produce relevance judgments. The Semantic Web community has used crowdsourcing for classic problems like ontology mapping where micro-tasks ask crowd workers to verify or identify mappings (e.g., 'is A a type of B'?) [26]. Crowdsourcing has also been used for ontology engineering in the biomedical domain showing that crowd workers, when provided with the relevant background knowledge, can perform as effectively as domain experts in this task [23].

5.2 Knowledge Graph Applications

When applied to Knowledge Graph problems, crowdsourcing has been used for entity linking [8], search query understanding [12], search result extraction [2], and Knowledge Graph enrichment [18].

ZenCrowd [8] combines both algorithmic and manual entity linking by dynamically assessing the quality of human work and aggregating crowd answers with algorithmic results based on a probabilistic reasoning framework. Input documents are first processed over a classic NLP pipeline (see Sect. 3) where entities are extracted by means of NER techniques and a ranked list of candidate entities from the Knowledge Graph is generated. At this point, a decision engine decides, based on entity linking result confidence, which entity to crowdsource. Once crowd answers are available, they are sent together with the original algorithmic results to a probabilistic network for the final linking decision. The probabilistic network combines prior probabilities form the entity linking method and crowd worker confidence scores and, by weighting in these signals, decides which links are to be selected for a certain entity. Experimental results have shown that (1) the use of crowdsourcing improves the quality of links generated by algorithmic approaches and that (2) the probabilistic network combination of human and machine answers improve the quality of the results over the plain crowd answers. This is possible thanks to the identification over multiple entities of trustworthy workers in the crowd and by giving an higher weight to their answer. This shows the importance of being able to identify the best workers in the crowd for a certain task which is an active area of research [6, 14].

The idea of using a probabilistic network to combine crowd and algorithmic results has been used also for data integration where the same entity appearing in multiple datasets has to be identified [9]. In this case a three-way blocking approach has been defines where crowdsourcing is seen as the most expensive similarity measure.

A final example of crowdsourcing applied to entity-oriented search is CrowdQ [12]: A system that uses crowdsourcing to understand the intended meaning of complex Web search queries by building a structured query template and answering it over a Knowledge Graph.

6 Open Research Problems

6.1 Knowledge Graphs

In the area of Knowledge Graphs there is a number of open research questions. One of these is about Knowledge Graph growth: As discussed in Sect. 2, Knowledge Graphs are often incomplete (e.g., because of the notability criteria of Wikipedia). This requires techniques to add relations to the graph (i.e., link prediction), to connect different graphs (i.e., ontology matching), and to add new entities to the Knowledge Graph. More than that, existing information may not be correct. Errors may be due, for example, to non-perfect entity resolution which causes the existence of duplicate entities in the Knowledge Graph. Another example of open research question is about how to best let users access data in the Knowledge Graph as we cannot expect the average Web user to type a SPARQL query. In this direction of research there is the need to work on *semantic parsing* to better interpret user queries and on *question answering* systems built on top of Knowledge Graphs.

6.2 Entity Search

Open research questions in the area of entity search go beyond the work on improving system effectiveness and efficiency for existing tasks. The heavy use of NLP pipelines requires a better understanding on how errors propagate over this pipelines and which effect they have on the final result presented to the end user. More than that, there is a need for work looking at the user experience and on novel entity-centric user interfaces that would allow user to access information in a richer way than just by using keywords to express an information need and consuming results as a ranked list. In this line of work, exploratory search of collections can benefit from entity-centric approaches allowing users to navigate hybrid graphs of documents and entities interconnected together.

Novel entity-centric search tasks can be looked at as well. Examples include searching for relations, attributes, or, in general, for more complex entity queries involving joins of datasets and complex requests (e.g., 'birthdate of the mayor of the capital city of Italy'). For such complex information needs the use of crowdsourcing has been proposed to solve the problem of query parsing and understanding [12,28].

6.3 Crowdsourcing

In Sect. 5 we have seen examples of so called hybrid human-machine systems that leverage machines to scale over large amounts of data as well as human intelligence by means of crowdsourcing to keep the quality of the results high.

While these systems generally provide more accurate results, a number of challenges have to be taken into account when designing such systems. First, the use of financial incentives attracts malicious workers who aim at obtaining the monetary reward attached to tasks without caring about the accurate completion of the task [17]. Moreover, the efficiency of hybrid systems is highly unpredictable because of the human-in-the-loop component which is difficult to schedule. A number of problems around effectiveness and efficiency of crowdsourcing platforms have still to be solved [7].

References

1. Balog, K., Fang, Y., de Rijke, M., Serdyukov, P., Si, L.: Expertise retrieval. Found. Trends Inf. Retrieval **6**(2–3), 127–256 (2012)
2. Bernstein, M.S., Teevan, J., Dumais, S., Liebling, D., Horvitz, E.: Direct answers for search queries in the long tail. In: Proceedings of the SIGCHI Conference on Human Factors in Computing Systems, CHI 2012, pp. 237–246. ACM, New York (2012)
3. Blanco, R., Cambazoglu, B.B., Mika, P., Torzec, N.: Entity recommendations in web search. In: Alani, H., Kagal, L., Fokoue, A., Groth, P., Biemann, C., Parreira, J.X., Aroyo, L., Noy, N., Welty, C., Janowicz, K. (eds.) ISWC 2013, Part II. LNCS, vol. 8219, pp. 33–48. Springer, Heidelberg (2013)
4. Blanco, R., Mika, P., Vigna, S.: Effective and efficient entity search in RDF data. In: Aroyo, L., Welty, C., Alani, H., Taylor, J., Bernstein, A., Kagal, L., Noy, N., Blomqvist, E. (eds.) ISWC 2011, Part I. LNCS, vol. 7031, pp. 83–97. Springer, Heidelberg (2011)
5. Blanco, R., Ottaviano, G., Meij, E.: Fast and space-efficient entity linking for queries. In: Proceedings of the Eighth ACM International Conference on Web Search and Data Mining, WSDM, Shanghai, China, 2–6 February 2015, pp. 179–188 (2015)
6. Bozzon, A., Brambilla, M., Ceri, S., Silvestri, M., Vesci, G.: Choosing the right crowd: expert finding in social networks. In: Proceedings of the 16th International Conference on Extending Database Technology, EDBT 2013, pp. 637–648. ACM, New York (2013)
7. Demartini, G.: Hybrid human-machine information systems: challenges and opportunities. Comput. Netw. **90**, 5–13 (2015)
8. Demartini, G., Difallah, D.E., Cudré-Mauroux, P.: ZenCrowd: leveraging probabilistic reasoning and crowdsourcing techniques for large-scale entity linking. In: Proceedings of the 21st International Conference on World Wide Web, WWW 2012, pp. 469–478. ACM, New York (2012)
9. Demartini, G., Difallah, D.E., Cudré-Mauroux, P.: Large-scale linked data integration using probabilistic reasoning and crowdsourcing. VLDB J. **22**(5), 665–687 (2013)
10. Demartini, G., Firan, C.S., Iofciu, T., Krestel, R., Nejdl, W.: Why finding entities in wikipedia is difficult, sometimes. Inf. Retr. **13**(5), 534–567 (2010)

11. Demartini, G., Missen, M.M.S., Blanco, R., Zaragoza, H.: TAER.: time-aware entity retrieval-exploiting the past to find relevant entities in news articles. In: Proceedings of the 19th ACM International Conference on Information and Knowledge Management, CIKM 2010, pp. 1517–1520. ACM, New York (2010)

12. Demartini, G., Trushkowsky, B., Kraska, T., Franklin, M.J.: CrowdQ: crowdsourced query understanding. In: CIDR, Sixth Biennial Conference on Innovative Data Systems Research, Asilomar, CA, USA, 6–9 January 2013, Online Proceedings (2013)

13. Difallah, D.E., Catasta, M., Demartini, G., Ipeirotis, P.G., Cudré-Mauroux, P.: The dynamics of micro-task crowdsourcing: the case of Amazon MTurk. In: Proceedings of the 24th International Conference on World Wide Web, WWW 2015, pp. 238–247. International World Wide Web Conferences Steering Committee, Republic and Canton of Geneva, Switzerland (2015)

14. Difallah, D.E., Demartini, G., Cudré-Mauroux, P.: Pick-a-crowd: tell me what you like, and i'll tell you what to do. In: Proceedings of the 22nd International Conference on World Wide Web, WWW 2013, pp. 367–374. International World Wide Web Conferences Steering Committee, Republic and Canton of Geneva, Switzerland (2013)

15. Dong, X., Gabrilovich, E., Heitz, G., Horn, W., Lao, N., Murphy, K., Strohmann, T., Sun, S., Zhang, W.: Knowledge vault: a web-scale approach to probabilistic knowledge fusion. In: Proceedings of the 20th ACM SIGKDD International Conference on Knowledge Discovery and Data Mining, KDD 2014, pp. 601–610. ACM, New York (2014)

16. Elmeleegy, H., Madhavan, J., Halevy, A.Y.: Harvesting relational tables from lists on the web. VLDB J. **20**(2), 209–226 (2011)

17. Gadiraju, U., Kawase, R., Dietze, S., Demartini, G.: Understanding malicious behavior in crowdsourcing platforms: the case of online surveys. In: Proceedings of the 33rd Annual ACM Conference on Human Factors in Computing Systems, CHI 2015, pp. 1631–1640. ACM, New York (2015)

18. Ipeirotis, P.G., Gabrilovich, E.: Quizz: targeted crowdsourcing with a billion (potential) users. In: Proceedings of the 23rd International Conference on World Wide Web, WWW 2014, pp. 143–154. ACM, New York (2014)

19. Li, C., Weng, J., He, Q., Yao, Y., Datta, A., Sun, A., Lee, B.-S.: TwiNER: named entity recognition in targeted Twitter stream. In: Proceedings of the 35th International ACM SIGIR Conference on Research and Development in Information Retrieval, SIGIR 2012, pp. 721–730. ACM, New York (2012)

20. Lin, T., Pantel, P., Gamon, M., Kannan, A., Fuxman, A.: Active objects: actions for entity-centric search. In: Proceedings of the 21st International Conference on World Wide Web, WWW 2012, pp. 589–598. ACM, New York (2012)

21. Macdonald, C., Ounis, I.: Voting for candidates: adapting data fusion techniques for an expert search task. In: Proceedings of the 15th ACM International Conference on Information and Knowledge Management, CIKM 2006, pp. 387–396. ACM, New York (2006)

22. Matuszek, C., Cabral, J., Witbrock, M.J., DeOliveira, J.: An introduction to the syntax, content of Cyc. In: AAAI Spring Symposium: Formalizing and Compiling Background Knowledge and Its Applications to Knowledge Representation and Question Answering, pp. 44–49 (2006)

23. Mortensen, J., Musen, M.A., Noy, N.F.: Crowdsourcing the verification of relationships in biomedical ontologies. In: AMIA, American Medical Informatics Association Annual Symposium, Washington, DC, USA, 16–20 November 2013 (2013)

24. Pound, J., Mika, P., Zaragoza, H.: Ad-hoc object retrieval in the web of data. In: Proceedings of the 19th International Conference on World Wide Web, WWW 2010, pp. 771–780. ACM, New York (2010)
25. Robertson, S.E., Zaragoza, H.: The probabilistic relevance framework: BM25 and beyond. Found. Trends Inf. Retrieval **3**(4), 333–389 (2009)
26. Sarasua, C., Simperl, E., Noy, N.F.: CROWDMAP: crowdsourcing ontology alignment with microtasks. In: Heflin, J., Sirin, E., Tudorache, T., Euzenat, J., Hauswirth, M., Parreira, J.X., Hendler, J., Schreiber, G., Bernstein, A., Blomqvist, E., Cudré-Mauroux, P. (eds.) ISWC 2012, Part I. LNCS, vol. 7649, pp. 525–541. Springer, Heidelberg (2012)
27. Smirnova, E., Balog, K.: A user-oriented model for expert finding. In: Clough, P., Foley, C., Gurrin, C., Jones, G.J.F., Kraaij, W., Lee, H., Mudoch, V. (eds.) ECIR 2011. LNCS, vol. 6611, pp. 580–592. Springer, Heidelberg (2011)
28. Teevan, J., Collins-Thompson, K., White, R.W., Dumais, S.: Slow search. Commun. ACM **57**(8), 36–38 (2014)
29. Tonon, A., Catasta, M., Demartini, G., Cudré-Mauroux, P., Aberer, K.: *TRank*: ranking entity types using the web of data. In: Alani, H., et al. (eds.) ISWC 2013, Part I. LNCS, vol. 8218, pp. 640–656. Springer, Heidelberg (2013)
30. Tonon, A., Demartini, G., Cudré-Mauroux, P.: Combining inverted indices and structured search for Ad-hoc object retrieval. In: The 35th International ACM SIGIR Conference on Research and Development in Information Retrieval, SIGIR 2012, Portland, OR, USA, 12–16 August 2012, pp. 125–134 (2012)
31. Zhiltsov, N., Kotov, A., Nikolaev, F.: Fielded sequential dependence model for Ad-hoc entity retrieval in the web of data. In: Proceedings of the 38th International ACM SIGIR Conference on Research and Development in Information Retrieval, SIGIR 2015, pp. 253–262. ACM, New York (2015)

A Short Survey on Online and Offline Methods for Search Quality Evaluation

Evangelos Kanoulas[(✉)]

Informatics Institute, University of Amsterdam,
Amsterdam, The Netherlands
e.kanoulas@uva.nl

Abstract. Evaluation has always been the cornerstone of scientific development. Scientists come up with hypotheses (models) to explain physical phenomena, and validate these models by comparing their output to observations in nature. A scientific field consists then merely by a collection of hypotheses that could not been disproved (yet) when compared to nature. Evaluation plays the exact key role in the field of information retrieval. Researchers and practitioners develop models to explain the relation between an information need expressed by a person and information contained in available resources, and test these models by comparing their outcomes to collections of observations.

This article is a short survey on methods, measures, and designs used in the field of Information Retrieval to evaluate the quality of search algorithms (aka the implementation of a model) against collections of observations. The phrase "search quality" has more than one interpretations, however here I will only discuss one of these interpretations, the effectiveness of a search algorithm to find the information requested by a user. There are two types of collections of observations used for the purpose of evaluation: (a) relevance annotations, and (b) observable user behaviour. I will call the evaluation framework based on the former a collection-based evaluation, while the one based on the latter an in-situ evaluation.

This survey is far from complete; it only presents my personal viewpoint on the recent developments in the field.

1 Introduction

The growth of the Web and the consequent need to organise and search a vast amount of information has demonstrated the importance of Information Retrieval (IR), while the success of web search engines has proven that IR can provide eminently valuable tools to manage large amounts of data. Evaluation has played a critical role in the success of IR. There is an arsenal of methods in hand that researcher and practitioners use to evaluate an experimental search system and compare it to the production system; in this paper we focus on the two predominant paradigms: collection-based evaluation and in-situ evaluation.

Collection-based evaluation is performed offline, in a laboratory setting. A test collection, comprising benchmark documents, queries, and human judgment

© Springer International Publishing Switzerland 2016
P. Braslavski et al. (Eds.): RuSSIR 2015, CCIS 573, pp. 38–87, 2016.
DOI: 10.1007/978-3-319-41718-9_3

labels of the relevance of each document to each query, together with an evaluation measure that summarises the relevance of a ranked list of documents returned
as a response to a query, are used to assess the effectiveness of a retrieval system [71,127].

On the other hand, in-situ evaluation is run online, by deploying an experimental system and running users queries both against the experimental and the
production system. A/B testing and interleaving provide between-subject and
within-subject experimental designs [41,98,139].

2   Setting the Stage

In a typical retrieval scenario a user, while performing a task, finds herself in
an anomalous state of knowledge [19]. Having access to an information retrieval
system the user searches for information useful to complete her task. Consider
the example in Fig. 1. The user is planning her holidays to Madrid, Spain and
she is interested in finding the current exhibitions at Prado museum. She poses
the query *Prado* to a search engine, the search engine accesses a collection of
searchable items - in this case an index of web documents - and returns a ranked

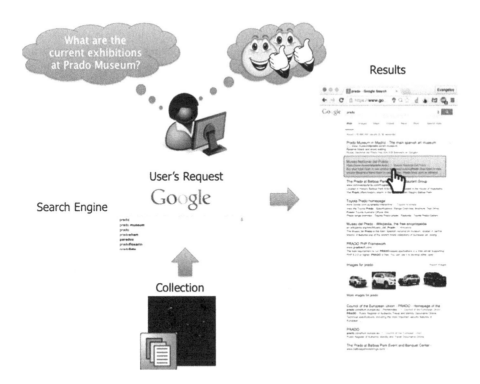

Fig. 1. A user searching and finding information about current exhibitions at Prado
museum.

list of documents to the user, through a search engine interface. The user then clicks on the first organic result (that is the first result after the advertisement), which corresponds to the official website of Prado Museum, navigates within the site and finds the information on current exhibitions. Hence, from the user's perspective, this first result was relevant to what she was looking for, and further useful towards the completion of her task, the visit to Madrid.

The goal of information retrieval evaluation is to quantify the user's satisfaction given the output of a retrieval system[1]. In an alternative formulation, given two or more search algorithms and their output to users requests, the goal of information retrieval evaluation is to quantify the relative difference in users' satisfaction from the output of the search algorithms.

Evaluation sits at the core of the scientific method for devising new laws, that is models of how the world functions. In hard sciences one is looking for a new law by first guessing it - making a hypothesis - then computing the consequences of this hypothesis, and last comparing the results of this computation to observation in nature. If the observations disagree with the experimental results then the hypothesis is wrong. If not, there is not enough evidence to reject the hypothesis, and this hypothesis becomes a new law, until future observations allows its rejection.

Information retrieval is no different to hard sciences in this respect. The goal of an information retrieval system is to match information seekers with the information they seek. Hence, research focuses on finding new laws, models that describe the relation between the information in a corpus and the information need of a user as expressed by her query. Computing the consequences of such a law corresponds to the algorithmic implementation of a retrieval model, and comparing to the observations in nature corresponds to running an information retrieval evaluation experiment. There is a variety of approaches to test whether the consequences of a hypothesis agree with observation, which typically take one of the following three forms: user studies, collection-based evaluation, and in-situ evaluation.

The three approaches have their pros and cons. We briefly discuss these along the following dimensions:

- **fidelity** of the method;
- **generalizability** of drawn conclusions;
- **cost** of the evaluation;
- **reusability** of the collected signals;
- **reproducibility** of the results.

In user studies human subjects directly interact with the search engines under evaluation in a laboratory setting and provide direct feedback. Given that no assumptions are made to interpret the user feedback, user studies allow the highest possible fidelity of the evaluation results. However, given that the number of human subjects (users) are typically small in these studies a careful experimental design is necessary to allow the generalisability of the conclusions drawn.

---

[1] Retrieval systems and search engines are used interchangeably in this paper.

For the evaluation of any new search algorithms a new user study is required, making user studies expensive. Furthermore, the results of a user study are hard to reproduce. In collection-based evaluation the user is abstracted out by the use of a test collection. Collection-based evaluation first makes a compositionality assumption: the overall users' satisfaction with the search engine's results can be decomposed to user satisfaction with the individual documents returned. The compositionality assumption is encoded by an evaluation measure. Further, given that user satisfaction is hard to measure it is replaced by topical relevance, i.e. how relevant a document is to a user's query. Collection-based evaluation provides the least fidelity since strong assumptions are made between relevance and user satisfaction. However, test collections can easily be reused in future experiments and the results of these experiments are reproducible. Furthermore, the generalisability of the conclusions drawn can be tested by statistical methods. Collecting relevance judgments is expensive; however methods like crowdsourcing [4,5,89,103], and statistical sampling [9,12,143,144] have been developed to reduce the cost. Further, the cost can be amortised across experiments given that the constructed collection can be used multiple times. In-situ evaluation sits in between the previous two approaches. It requires a search algorithm to be deployed online and used by real users. This already provides higher fidelity compared to the collection-based evaluation since queries come directly from the users of a search engine. Even though user satisfaction does not need to be replaced by relevance, typically users do not provide direct feedback. Thus, there is an assumption made that user satisfaction can be inferred by online observable user behaviour. Evaluation measures then are applied over observable user behaviour to infer user satisfaction. Experiments are not easily reproducible, and the test collection generated (query logs) are not easily reusable.

This work will focus on collection-based and in-situ evaluation, discussing recent work and open problems.

2.1 Collection-based Evaluation

Test collections in information retrieval are similar to test collections used in other fields of computer science, such as machine learning or computer vision. Figure 2 demonstrates differences and similarities. In Machine Learning a test collection consists of a set of <feature vector, label> vectors. Features are given to a machine learning algorithm which is evaluated with respect to how accurately it can predict the labels. Somewhat differently, in computer vision, it is the raw images, i.e. the recognisable items that are provided instead of features, together with labels. A computer vision algorithm then extracts features from these images and use these features towards predicting the label of each image. Information retrieval test collections bear stronger similarities to computer vision collections than to machine learning ones in the sense that it is <query, document, label> vectors that are provided as a test collection, instead of <feature vector, label> vectors. A search algorithm then extracts features

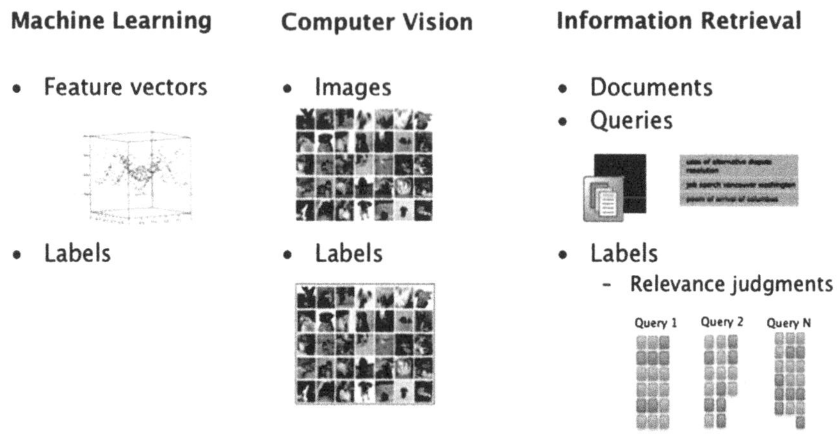

Machine Learning   **Computer Vision**   **Information Retrieval**

- Feature vectors   • Images   • Documents
                                                • Queries

- Labels   • Labels   • Labels
                                                  – Relevance judgments

                                                  Query 1   Query 2   Query N

Fig. 2. A static test collection.

from query-document pairs and tries to predict the label of that pair. In information retrieval, the task is actually easier than trying to predict the exact label of the pair. Given that the output of a search engine is a ranked list of documents, it is only the relative order of <query, document> vectors for a given query that really matters.

The question that arises is how can we build such a test collection? Ideally, one would like to label all documents in the collection against all queries, however this is rather expensive; hence we can only label a sample of them. In most machine learning tasks one draws a uniform random sample from the universe of all items, and annotates this sample with labels. This is not a good option for information retrieval due to the skewness of the data. That is, for a given query, only a tiny percentage of documents are relevant, and hence will have a positive label, while most of the documents are irrelevant. Figure 3 demonstrates how test collections are typically constructed. Instead of sampling the universe of all documents in the collection multiple search algorithms are used to retrieve documents for a given query. The ranking of documents on the basis of all these algorithms is then considered in the selection process of labelling documents. TREC[2] has established a depth-k pooling method, in which, a pool of documents that appear in the top-k of any ranking is only considered and labeled. The remaining of the documents in the collection are considered irrelevant under the assumption that if none of the search algorithms pooled retrieves these documents at the top-k ranks then it is very unlikely that these documents are relevant. There is a large volume of research looking whether this assumption holds, and whether it affects the conclusions of an experiment.

---

[2] Text REtrieval Conference.

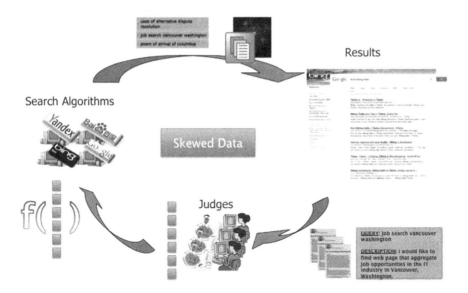

Fig. 3. Static test collection construction methodology.

After documents are selected they are handed to human assessors who provide a relevance judgment for each document in the pool against the test query. To evaluate a search engine then one replaces the ranked list of documents retrieved by the corresponding ranked list of relevance judgments and an evaluation measure aggregates the relevance judgments of the list towards a single value that expresses the quality of the search engine. Clearly, a measure values should align as much as possible to user satisfaction.

Almost all research questions concerning collection-based evaluation can be expressed with a reference to Fig. 3. Here are some questions that IR researchers have been investigating.

1. How should we choose benchmark queries and document to test search algorithms on, and how many of them is necessary and sufficient for reliable and generalisable conclusions? (See [20, 21, 66, 77–79, 110, 121, 151].)
2. How should we choose documents to judge and how many of them is necessary and sufficient for reliable and generalisable conclusions? (See [3, 9, 12, 24, 29] and [31, 34–36, 143, 144].)
3. Who are the judges and how many relevance labels shall we collection per query-document pair? (See [2, 4, 5, 14, 22, 37, 78, 89–91, 103, 129, 130, 145, 146, 148].)
4. Do judges agree with each other and what is the impact of their disagreement on retrieval evaluation? Do they agree with the actual users of a search engine? (See [39, 54, 92, 109, 111, 141].)

5. Shall we provide a description of the query intent to judges [125, 145]?
6. Is relevance binary, and if not then what [108, 140]?
7. What is a good evaluation measure to aggregate relevance judgments?
8. How do we determine what a good measure is? How do we evaluate evaluation measures?
9. How can we compare two or more search engines on the basis of these evaluation measures (comparative evaluation)?

The last three questions will be further explored later in this article.

2.2   In-situ Evaluation

In-situ evaluation requires a search algorithm to be deployed online. Users can then interact with the live system and provide feedback regarding the quality of the search results.

Given that search engine users are typically reluctant to provide explicit feedback the evaluation infrastructure logs users interacting with the live system under the assumption that the quality of a search engine can be inferred by these interactions. Implicit in this assumption is that users behave rationally: they have a goal when they use the search engine and consistently work towards that goal. Therefore, they are not submitting random queries or click random results and they do not provide malicious data to the system.

Logged interactions (observable user behaviour) include clicks on search results, dwell time spent on the SERP or at the landing pages (clicked results), mouse movements, browser actions (bookmarks, saves, prints), and query reformulations. Given this implicit feedback the goal of an in-situ evaluation methods is to infer the quality of the search engine under use. Figure 4 shows a sample of logged interactions in the AOL query log.

The biggest challenge in in-situ evaluation is the interpretation of the observable user behaviour towards quantifying the quality of a search algorithm. This requires a careful experimental design. Decision regarding the design are made along two dimensions:

- Document level vs. SERP level interpretation;
- Absolute vs. relative interpretation

In former case the design decision relates to the level at which user observable behaviour needs to be interpreted: do we want to infer the quality of documents constituting a SERP or the overall quality of the SERP? In the latter case the design decision relates to inferring the absolute quality of a document/SERP or the relative quality between two or more documents/SERPs. Based on these two dimensions a number of evaluation methods that interpret user observable behaviour have been devised. Examples of those can be seen in Table 1.

Fig. 4. A sample from the AOL query log.

Table 1. Classification of in-situ evaluation measures/methods.

Evaluation method	Absolute	Relative
Item level	Click-through rate, ...	Click-skip, ...
SERP level	Abandonment rate, ...	A/B testing, interleaving, ...

There two predominant experimental designs used for inferring search quality by observable user behaviour are:

- **A/B testing**. An percentage of query traffic uses system A (baseline or control system) while the remaining of query traffic use system B (experimental or treatment system). This is a between-subject experiment since different queries are handled by different systems.
- **Interleaving**. A combination of search engine results from system A and system B are shown to the user. This is a within-subject experiment since the same query is handled by both systems.

An example of an A/B testing experiment can be seen in Fig. 5. Two different versions of the same search engine are deployed. Some query traffic is handled by the baseline system, while the remaining traffic is handled by the experimental system. Users to don't explicitly provide feedback about the quality of the search results they receive. Instead, as discussed earlier, implicit feedback is used to infer the quality difference between the two systems.

Baseline (control)   **Experimental (treatment)**

Fig. 5. A/B Testing.

In-situ evaluation provides high fidelity: real users replace the judges; there is no ambiguity in their information need; users actually want results; and performance is measured on real queries.

3   Collection-based Evaluation

In this section we will focus on two significant issues regarding collection-based evaluation: (a) how to collect relevance judgments to include in the benchmark test collection, and (b) how to aggregate these judgments to evaluate the quality of a search engine.

3.1   Obtaining Relevance Judgments

Obtaining relevance judgments is expensive; it requires human annotators to manually assess the quality of documents in the collection against user queries. The overall cost is equal to the *cost per annotation* × the *number of annotations* necessary to evaluate the quality of a search engine.

Crowdsourcing[3] has been used to reduce the cost per annotation by hiring laypeople to judge documents. These annotations are often noisy or erroneous and most of the research in this area has focused on distilling the signal in the annotations from the noise [4,5,89,103].

---

[3] See the TREC Crowdsourcing track: https://sites.google.com/site/treccrowd/.

On the other hand, one may try to reduce the annotations in the benchmark collection to reduce the cost. The question that arises is how many judgments are necessary for the *reliable* evaluation of tested search engines. There is a rich literature in identifying this number, as well as how to effectively choose the documents to be annotated. Two significant factors affect the answers to the aforementioned questions:

- **[Recall]** There are information seeking tasks (e.g. patent search, legal search, systematic reviews, etc.) for which the entire set of relevant documents in the collection need to be retrieved; the evaluation of search engines on these task requires that all relevant documents identified are identified and annotated.
- **[Reusability]** The constructed benchmark collection is typically not only used to measure the quality of the search engines under evaluation, but it is used to test future systems; this requires again that all relevant documents in the collection are identified and annotated so that if a future system brings up a new relevant document that has not been encountered before by any system that was used to assemble the test collection the evaluation measure awards this system.

Therefore, ideally one would like to annotate all documents in the corpus against user queries. Given that this is practically impossible two different approaches have appeared in the literature to select only a subset of documents to be annotated and still enable recall-based evaluation and reusability of the benchmark collection: (a) deterministic approaches, and (b) stochastic approaches.

The former deterministically choose documents to be annotated that fulfil certain criteria, e.g. the have a high probability of being relevant, or they are the most discriminative in comparing two search algorithms. The latter uses sampling methods to select a number of documents to be annotated, and then infer the quality of the search algorithms (or the relevance of the rest of the unjudged documents) using this sample.

Deterministic Approaches: Some of the deterministic approaches for selecting documents to be annotated are the following:

- **Depth-k pooling:** The top-k results from the search engines currently under evaluation are pooled together and annotated by human annotators. Documents that do not appear in the top-k of any system under evaluation, and hence not annotated, are considered irrelevant.
- **Automatic evaluation:** Evaluation is performed without annotations; no relevance judgments are obtained. Instead the "majority vote" regarding the relevance of a document to a query obtained by the different search engines in this experiment is considered as a signal of relevance and it is used in the evaluation [61,73,113,137,142]. Automatic evaluation methods have been shown to suffer from the "tyranny of the masses" [10] with the best systems typically retrieving good but rare documents, which are not retrieved by the

majority of the systems, and hence considered irrelevant, leading to an under-estimation of the quality of these systems.

- **Meta search:** The participating in the experiment search engines are collectively used to identify the most relevant documents and those documents are prioritised towards being annotated. Some of the algorithms in this category are:
  - Move-to-Front pooling [48]
  - Interactive searching and judging [48]
  - Hedge [8]
- **Minimum Test Collection (MTC):** Documents are selected based on how well they can discriminate systems participating in the experiment based on their quality [24,29,31,34–36]. If for instance two systems retrieve the same documents at the top-3 positions of the returned ranked list, these documents are the least discriminative, since they cannot tell apart the two systems.
- **Nugget-based evaluation:** The annotation of relevance is done on the passage level rather than the document level. Relevant passages are then used to identify close matches of passages in unjudged documents and propagate relevance [114].

Stochastic Approaches: A different approach in selecting documents to judgment is inspired by statistical sampling and inference methods. Stochastic methods do not intent to identify all relevant documents, or documents that can tell apart two systems, but instead they are designed to estimate the quality of a search algorithms using a small number of randomly selected documents [9,143]; see Fig. 6. The methods developed work as follows; first the measure of interest is defined as the outcome of a random experiment; then it is estimated using random sampling. As an illustrative example let's consider the precision in the top-10 retrieved documents by a search engine, $P@10 = \frac{\sum_{i=1}^{10} \mathbb{I}(d_i \in \mathcal{R})}{10}$. Precision at cut-off 10 can be defined as the expected outcome of a two step experiment:

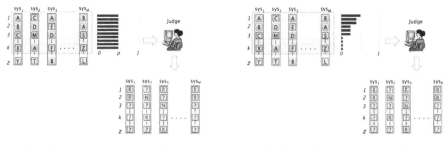

(a) Uniform random sampling   (b) Stratified random sampling

Fig. 6. Sampling methods for selecting documents to be judged as an alternative to depth-k pooling.

(1) Select a rank at random from the set {1,...,10}, and (2) Output the binary relevance of document at this rank. Another example is average precision,

$$AP = \frac{\sum_{r \in \mathcal{R}} \frac{\sum_{i=1}^{r} \mathbb{I}(d_i \in \mathcal{R})}{i}}{|\mathcal{R}|}$$

Average precision can also be defined as the expected outcome of a three step experiment: (1) Select a relevant document at random; let the rank of that document be k, (2) Select a rank at random from the set {1,...,k}, and (3) Output the binary relevance of document at this rank. On the basis of the afore-described definitions one can randomly sample from the pool of documents constructed by all search systems in the experiment and calculate these measures. Other measures can be estimated in the same fashion. The estimated values of the measures are by construction unbiased, however they vary across different samples. The variance can be mathematically derived [144]. To reduce variance, given that most evaluation measures give more weight to documents towards the top of the ranked list, "top-heavy" sampling strategies based on importance and stratified sampling have been deployed [144]. Finally, using the estimated value of evaluation measures Aslam et al. [11,12] uses constraint optimisation to estimate the relevance of the unjudged documents. Constraints come from the fact that different search engines may be able to retrieve the same unjudged documents at different ranks. The relevance of these documents are estimated so that it agrees across the different search engine ranked lists, and the measure calculated with the estimated relevance of these documents agrees with the estimated measure using the initial sample and the provided annotations.

3.2   Evaluation Measures

Evaluation measures aggregate the relevance judgments over a ranked list of returned documents to determine the quality of a search engine. Traditional evaluation measures, such as precision, recall, and average precision, essentially extend set-based measures to ranking measures and capture how good is a system in identifying relevant documents. However, a good measure of retrieval effectiveness should also correlate with the users experience when using a search engine. To align evaluation measures with user experience researchers developed models of user behaviour (interactions of users with the returned search results) and defined evaluation measures based on these models.

Model-based Evaluation Measures: A model-based evaluation measure essentially depends on three separate models [25]:

- a browsing model that describes how a user interacts with results;
- a model of document utility, describing how a user derives utility from individual relevant documents;
- a utility accumulation model that describes how a user accumulates utility in the course of browsing.

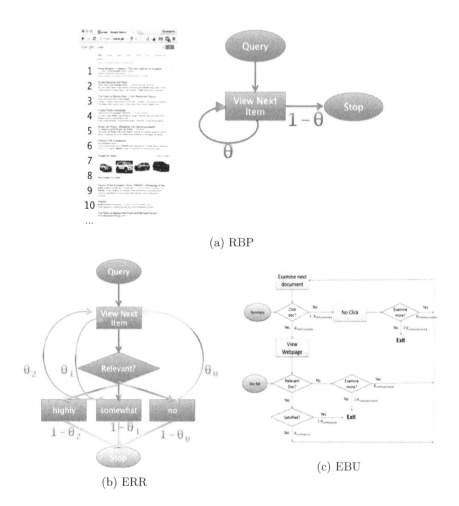

(a) RBP

(b) ERR

(c) EBU

Fig. 7. User models

Most of the research has focused on the browsing and document utility model. Regarding the former, the *position-based model* assumes that the chance of observing a document only depends on the position of the document in the ranked list [112]; while the *cascade-based model* also considers the interactions of the user with documents higher in the ranked list than the document under consideration [42,50,147]. Figure 7 show the user model for three popular model-based evaluation measures: Rank-Biased Precision (RBP) [112], (b) Expected Reciprocal Rank (ERR) [42], and (c) Expected Browsing Utility (EBU) [147].

Chuklin et al. [43,45] considered models developed to predict on which documents a user may click while browsing a search engine result page (SERP), and and demonstrated how one can develop an arsenal of evaluation measures based on these click models.

Regarding the document utility model, Smucker and others [62,136] considered time as part of the effort a user makes into finding relevant information; hence the utility of documents low in the ranked list or long documents is penalised.

Novelty and Diversity-based Evaluation Measures: Queries are inherently ambiguous or faceted. An automatic system can never know the users intent. Diversification attempts to retrieve results that may be relevant to a space of possible. The evaluation measures mentioned so far however to do not consider the possible multiple intents behind a query. Instead they assume that there is a single universal intent behind any user query. Annotations are made based on this unique intent. Accounting for multiple possible query intents requires (a) identifying these intents prior to constructing the benchmark collection, (b) annotating documents against each one of the identified intents, and (c) developing measures that account for the relevance of a document to the different intents.

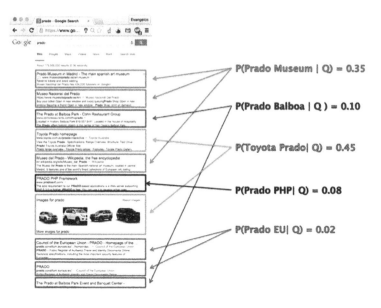

Fig. 8. Query intents for the query *Prado*

Intent-Aware measures. [1] assume there is a probability distribution $P(i|Q)$ over intents for a query Q; as an example see Fig. 8. This is the probability that a randomly-sampled user has the intent i when submitting query Q. Then the intent-aware version of a measure is its weighted average over this distribution.

Intent-aware measures award systems that return documents which can collectively cover multiple intents. However, returning 10 documents relevant to a single intent yields the same system performance as returning 10 documents relevant to 10 different intents. This is due to the fact that intent-aware measures do not penalise redundant information, that is documents that are relevant to the exact same query intent. α-nDCG is a generalisation of nDCG that accounts for both novelty and diversity [40,47]. The parameter α is a geometric penalisation for redundancy. The measure redefines the utility of a document as follows:

- +1 for each intent it is relevant to
- $(1-)$ for each document higher in the ranking that intent already appeared in

One of the key questions in constructing a novelty and diversity benchmark collection is how to identify the possible query intents. Radlinski et al. [49,120] devised an algorithm based on query reformulations and the click graph to infer query intents. The algorithm has three phases: The *Expand* phase finds the $k = 10$ most frequent valid reformulations of q, then the k most frequent valid reformulations of those. q' is a valid reformulation of q if (1) q' was followed by q within ten minutes by at least 2 distinct users, and (2) of all pairs of queries (q_i, q') issued by any user within 10 min, (q, q') occurred at least a fraction δ of the time (with δ set 0.001); the *Filter* phase reduces the query neighbourhood to more closely related queries, improving precision. We connect two queries if they were often clicked for the same documents, using a two step random walk on the bipartite query-document click graph. All pairs of queries with a random walk similarity above a fixed threshold are connected (this may add links not present in the reformulation graph, and usually removes many others). Additionally, all components of size less than t are removed completely (with t set to 2); the *Cluster* phase uses the random walk similarities to find intent clusters. Other algorithms, based on user behavioural data and/or crowdsourcing, have appeared in the effort to discover query intents based [52,80,91,116], while work has also be done in identifying what is a representative sample of query intents for reliable evaluation [145].

Session-based Evaluation Measures: The afore-described evaluation framework focus on serving the best results for a one-shot query. However, users frequently reformulate their initial query. The questions that arise are: Can we measure the effectiveness of a retrieval system over a sequence of reformulations (session)? Can we optimise systems to provide better results over a session? The TREC Session Track[4] was proposed and funded to extend the evaluation framework from one query evaluation to multi-query session evaluation by (a) construct the necessary benchmark collections, and (b) establish new evaluation measures.

---

[4] http://ir.cis.udel.edu/sessions/.

(a) Construction time.

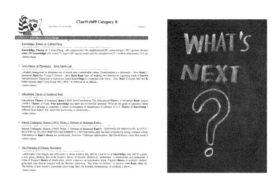

(b) Testing time.

Fig. 9. Session-based collection

Constructing a session-based benchmark collection remains an open problem. Figure 9 demonstrates the key difficulty. Figure 9(a) demonstrates the collection construction phase. A search engine is used to collect queries and reformulations, while the returned results are annotated similarly to the traditional benchmark collection. However, at testing time, shown in Fig. 9(b), given the first query in the multi-query session collected before, the tested system reacts in a different manner than the system used to create the collection, returning different documents. Given this new ranked list of documents it is an unrealistic assumption to consider that the user will reformulate her query the same way as before; hence there is no way to know how to evaluate the next step in the session. Simulating users instead of real users has appeared in the literature as a solution to this problem [17, 18, 30, 83].

Given a (possibly simulated) session-based benchmark collection, one also needs to rethink of the evaluation measures that can capture the goodness of a system across the entire user session, and not just for a single query. Measures that have appeared in the literature typically extend user models constructed for a single query scenario to the multi-query case. They assume that the user steps down a ranked list of documents, observes each one of them until a decision point and either (a) abandons the search, or (b) reformulates. While stepping down or sideways, the user accumulates utility. In the case of the Session DCG [82] the user steps down the ranked list until rank k and always reformulates. This is a deterministic model which does not allow early abandonment. Yang and Lad took a stochastic approach where the user steps down the ranked list of results, one-by-one, and stops browsing documents based on a stochastic process that defines a stopping probability distribution over ranks and reformulates. Their model is stochastic but it also does not allow early abandonment; that is the user will continue reformulating as long as there are reformulated queries in the collection. Last, Kanoulas et al. [88] proposed a number of expected session measures, according to which the user steps down a ranked list of documents until a decision point and either abandons the query or reformulates.

Evaluation of Evaluation Measures: Evaluation measures are of supreme importance to the development of search engines. Search algorithms are optimised with respect to some evaluation measure. Thus, it is the measure that actually defines the problem that is being solved. Tens of search quality measures have appeared in the literature to accommodate the different information access tasks researchers and practitioners study [53]. The question that arises is what makes a good measure, and how to select or define one. Two frameworks have been used to evaluate evaluation measures, (a) an axiomatic definition framework, and (b) an empirical evaluation framework.

The former defines a set of axioms (or constraints) that an evaluation measure should follow. These axioms are derived from the information access task under study. Measures are then mathematically derived from these axioms or tested against them [6,23].

For the empirical evaluation a number of methods have been developed that either compare the outcome of an experiment (as dictated by the evaluation measure value) with implicit or explicit user feedback, or they examine inherent properties that good measures should demonstrate when used for evaluation.

Query Logs (User): Modern evaluation measures are typically defined on the basis of some model of user interaction with the SERP. Instead of directly evaluating the measure one can evaluate the user model, by comparing the predictions that the model makes to the data observed in query logs. Log-likelihood and perplexity are used to essentially measure the goodness-of-fit of a model to the clicks observed in a log [42,44,64].

Side-by-Side and Click-based (User): Side-by-side experiments, in which the results of two search algorithms under evaluation are shown to a user and the user expresses her preference can be used as a different evaluation framework. Click-based measures (some of which will defined later in this paper) can also aggregate online clicks to compare two search algorithms. A way to evaluate an evaluation measure is then to measure the agreement of the conclusions drawn by using this measure and a side-by-side or click-based measure regarding which algorithms is better [128].

Discriminative Power (Property): Evaluation measures are typically used in a comparative setup where two or more algorithms need to be compared with each other. As we will see in the next section of this paper, statistical significant test are used to allow the inference of the comparisons from a sample of test queries to the population of all possible user queries. An evaluation measure defines the power of such a test, that is its ability to tell that one system outperforms another. Hence, two evaluation measures can be compared with respect to their discriminative power [123, 126].

Informativeness (Property): A search engine performance is defined by the relevance of the results page. Any evaluation measure receives as an input this ranked list of relevance judgments and aggregate it to summarises the performance of the search engine. Hence, the only information that an optimisation process have regarding the quality of the SERP is though the value of an evaluation measure. The more informative a measure is the less uncertainty there is regarding the ordering of relevance judgments it has summarised. Using an informative measure versus an uninformative one is particularly important when evaluation measures are used as objective functions in a learning-to-rank framework. Hence, one can evaluate a measure on the basis of its informativeness. Aslam et al. has used a maximum-entropy framework to find the most likely distribution of relevance over a ranked list given a measure value [13]. The formulation of the maximum entropy framework was then extended by Ashkan for user-based and novelty and diversity-based measures [7].

4   Statistical Significance Testing

4.1   Hypothesis Testing 101

Let's assume now that we have carefully chosen the measure to quantify the quality of the retrieval algorithms we want to test. Further, let's also assume that we have a hypothesis we want to test: "if we rewrite a user's query by including all the synonyms of the terms in the query the effectiveness of a query likelihood ranking algorithm, as measured by average precision, will increase".

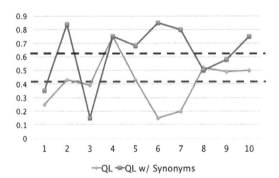

Fig. 10. Query likelihood with and without query expansion.

We select a set of queries to test the two ranking algorithms: (a) the baseline: a query likelihood ranking algorithm, and (b) the experimental: a query likelihood ranking algorithm with query expansion by synonyms. We obtain relevance judgments for the query-document pairs in our collection and we measure the effectiveness of your system by the chosen measure, e.g. Average Precision (AP). The results of the query likelihood system with and without query expansion and be seen in Fig. 10. If we look at the two plots, what can we say about our hypothesis? Looking at the mean AP values for the two systems one may be inclined to conclude that our hypothesis is correct. However, is it possible that the observed results are due to randomness in the choice of the 10 queries? That is, will the results hold if we throw away these test queries and choose another set of test queries; or in general do the results from these 10 queries generalise to the population of all queries?

Statistical significance testing[5] allows to test whether the difference in the measurements between the two systems are due to randomness in the sample of queries and hence they cannot be generalised to the population of all queries.

To perform a statistical significance test first we form the *null hypothesis*. This is the hypothesis we want to prove wrong, hence it is the opposite of our working hypothesis.

- $H_0 : \mu_A - \mu_B = 0$
- $H_a : \mu_A - \mu_B \neq 0$ or $\mu_A - \mu_B > 0$

Then we obtain system performance measurements over a sample of queries, and compute a *test statistic* t from those measurements. The statistic t should be chosen so that it has a known distribution under H_0. Finding such a statistic is often difficult, and different tests differ exactly in the statistic they compute and the distribution this statistic follows under H_0. Having chosen the statistic t with a known distribution under H_0, and computed its value based on the

---

[5] A tutorial on the topic has also been given by Carterette [27,28].

measurements from the experiment, we can then calculate the *p-value*, the probability of observing this value for t while assuming that H_0 is true. If the p-value is very low, one can conclude that H_0 is false. If the p-value is not low then there is no evidence to allow the rejection of the null hypothesis. To make such a decision we need a threshold α over p-values. This threshold is typically set to 5 % (or 1 %).

The most commonly used tests in information retrieval are:

- Parametric
  - Students t-test
- Non-parametric
  - Sign test/binomial test
  - Wilcoxon signed rank test
- Distribution-free
  - Randomisation test
  - Bootstrap test

Parametric tests rely on assumptions about the shape of the distribution (e.g., assume a normal distribution) in the underlying population and about the form or parameters (i.e., means and standard deviations) of the assumed distribution. Nonparametric tests rely on no (or few) assumptions about the shape or parameters of the population distribution from which the sample was drawn, while distribution-free tests make no assumption about the underlying distribution what so ever. A comparison among the aforementioned tests can be found in Smucker et al. [134,135].

Let's take the Student's t-test as an example of a statistical significance test. The t statistic is defined as,

$$t = \frac{\mu_{B-A}}{\frac{\sigma_{B-A}}{\sqrt{N}}} \sim \mathcal{N}(0,1)$$

where μ_{B-A} and σ_{B-A} is the mean and the variance of the differences between the values of the two systems. Note that a paired t-test is used to compare two population means since we have two samples in which observations in one sample can be paired with observations in the other sample, given that the two systems run over the same queries and against the same collection.

Table 2 illustrates an example, where the two systems, A (Query Likelihood) and B (Query Likelihood with Synonyms) are ran against 10 queries. Each number in column A and B corresponds to the effectiveness of the two systems over a query, while the last column corresponds to the difference in the performance measurement. The t statistics computed on the basis of these numbers is 2.33.

Table 2. Average precision values over 10 queries for query likelihood with (system A) and without (system B) query expansion with synonyms. (These are hand-picked values, not the output of an actual experiment.)

Query	A	B	B − A
1	0.25	0.35	+0.10
2	0.43	0.84	+0.41
3	0.39	0.15	−0.24
4	0.75	0.75	0
5	0.43	0.68	+0.25
6	0.15	0.85	+0.70
7	0.20	0.80	+0.60
8	0.52	0.50	−0.02
9	0.49	0.58	+0.09
10	0.50	0.75	+0.025

Figure 11 depicts the distribution of the t statistics under h_0, that is under the hypothesis that there is no difference in the effectiveness of the two systems in the population of queries. The statistic computed based on Table 2 lies at the right tail of the distribution. The blue area, to the right of the computed t statistic, is the probability of observing a t value that is equal or greater to 2.33 under the null hypothesis, which can be computed to 0.0225. Given that this probability is smaller than 0.05 we can reject the null hypothesis, and conclude that including synonyms in the query improves the effectiveness of query likelihood.

The question that arises is whether there is any chance that the null hypothesis is still true while we reject it (Type I error), and what is this probability. Figure 12(a) illustrates this exact probability. Given that we reject the null hypothesis when the p-value of our test is less than or equal to 0.05, this is exactly the chance of making a mistake.

The same question holds in the case we could not reject the null hypothesis, that is, is there a chance that the null hypothesis is false, while we do not reject it (Type II error), and what is this probability. Answering this question is not as easy as the previous question. The reason is that to answer this question, one needs to know the distribution from which the observed t statistic comes if not from the one under the null hypothesis. However, this is something we cannot know. For the shake of the discussion however let's assume that we know this distribution. Figure 12(b) depicts both the distribution under the null hypothesis and the actual distribution which produces the observed statistic. Given that we cannot reject the null hypothesis for any value of t for which the p-value is greater than 0.05, the probability that we erroneously do not reject the null hypothesis under our working assumption that the observed t statistic comes from the second distribution in the plot is the red area.

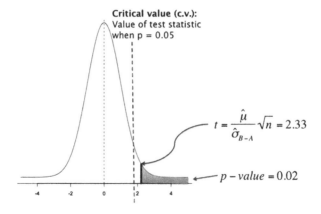

Fig. 11. The t statistic distribution under H_0 and the probability of the difference coming from this distribution. (Color figure online)

Statistical significance tests are classifiers. They predict whether a hypothesis is true or false. Hence one could construct a contingency table for a significance test (see Table 3).

To summarise Table 3 the test parameter α is used to decide whether to reject H_0 or not if $p < \alpha$, then we reject H_0. Choosing α is equivalent to stating an expected Type I error rate. E.g. if $p < 0.05$ we are saying that we expect that we will incorrectly reject H_0 5 % of the time. When H_0 is true, every p-value is equally likely to be observed, and hence 5 % of the time we will observe a p-value less than 0.05 and therefore there is a 5 % Type I error rate.

The power of statistical significance test is the probability the test correctly rejects the null hypothesis (1 - Type II error). Given that there are many ways the null hypothesis can be false we cannot simply compute the power of the test. However, we can set the minimum difference in the t statistic that the test should be able to detect. Setting this δ essentially sets the distribution under the

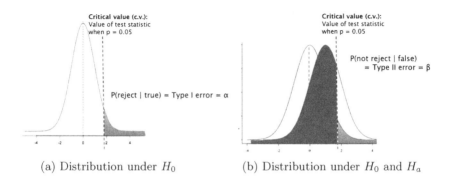

(a) Distribution under H_0 (b) Distribution under H_0 and H_a

Fig. 12. Student's t distribution. (Color figure online)

Table 3. Significance testing contingency table.

H_0	True	False
Not rejected	Accuracy: 1- α	Type II error: β
Rejected	Type I error: α	Power: 1 - β

alternative hypothesis. The t statistic is a function of the *effect size*, $\frac{\mu_{B-A}}{\sigma_{B-A}}$, and the size of the query sample, n. The effect size is a measure of the magnitude of the difference between two systems, and it is dimensionless; intuitively it is similar to % change in performance. The bigger the population effect size the more likely to find a significant difference in a sample.

Before testing, we can say I want to be able to detect an effect size of h with probability β, e.g. If there is at least a 5 % difference, the test should say the difference is significant with 80 % probability. Once we have chosen α, β, and h, we can determine the sample size needed to make the error rates come out as desired: $n = f(\alpha, \beta, h)$

To summarise power analysis and statistical significance testing are the two sides of the same coin. When setting up an experiment one can decide the Type I and Type II errors, that is the probability to find an effect that is not there, and the probability of not finding an effect that is there. Further one can decide the mean difference in the measurement of the performance of two systems one would like to detect. Given that the effect size is defined as the mean performance difference divided by the standard deviation of the performance difference to identify the sample size we only need to know the standard deviation of the performance difference.

4.2 Significance Testing in IR

In collection-based evaluation of search engines the only source of variance considered and account for by the statistical significance test mentioned earlier is variance due to the random selection of queries. However randomness in the experiments come from a number of sources (Fig. 13):

- Properties of queries
- Properties of document corpus
- Properties of effectiveness measures
- Assessor disagreement
- Missing relevance judgments
- Etc.

Current practise in information retrieval experimentation only accounts for the variance due to sampling queries. All other sources of variance are ignored either by being conflated in the variance due to queries or ignored all together, which can lead to drawing wrong conclusions [26,32,33,122,124,145].

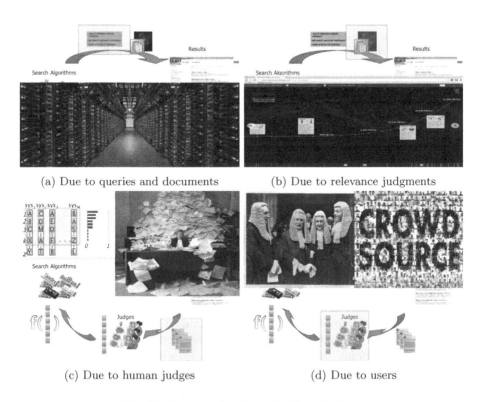

(a) Due to queries and documents   (b) Due to relevance judgments

(c) Due to human judges   (d) Due to users

Fig. 13. Sources of variance in IR evaluation.

In what follows we will examine, as an illustrative example, two additional to the queries sources of variance: (a) variance due to the properties of the document corpus, and (b) variance due to the behavioural properties of users.

When considering sources of variance other than the queries there are two questions that one needs to answer:

1. How can we quantify this variance?
2. How can we account extend the statistical test to account for multiple sources of variance?

The usual approach to system evaluation in IR experiments is to choose a measure defined on the results of searching on an individual query, and take an average of this measure over the set of queries, for the system concerned. The usual approach to statistical significance is to consider the results for the individual query as the units of measurement. That is, the fact that system A performs better than system B on topic 1 counts as a definitive result. The significance question arises only at the aggregation-over-topics level (does system A perform better than system B on significantly more topics?). We do not ask the question as to whether or in what sense the individual topic 1 result was significant. This approach effectively assumes that the topics were sampled from some population of (possible or actual) topics, but that the document collection is fixed in stone for all time.

Stated like this, it must be clear that this (the document collection part) is a rather drastic assumption. In fact it is in part mitigated by the general view in IR research that a result is only good if it works on multiple test collections. But this is a rather crude approach to the problem. It would be good to have a much better understanding of what each individual topic result is telling us, and of how best to draw general conclusions from the topic sample × document sample results that we have.

If it is accepted that the document collection should be regarded as sampled in some way from a population each per-topic measurement has some builtin error of estimation because the measurement is based on the sample rather than the population. To quantify the variance of a per-topic measurement across multiple samples of documents one can actually consider a document corpus and create equal size samples from it. A different approach used in Robertson and Kanoulas [122] is to simulate different corpora. There are different approaches to that: sampling the collection with replacement, bootstrap sampling of the document scores computed by a search algorithm, add noise to the bootstrapped scores (Kernel Density Estimation), or carefully fit a smooth model to the scores and sample from this model. In all cases the evaluation measure can now be computed for the different samples of the corpus and the effectiveness of the algorithm does not only vary across queries but also within queries (see Fig. 14).

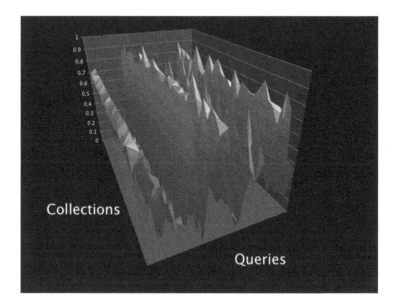

Fig. 14. Search algorithm performance per query and document corpus.

The t-test is based on a linear regression model that can only account for a single source of variance,

$$y_{ij} = \beta_i + b_j + \epsilon_{ij}$$

$$b_j \sim \mathcal{N}(0, \sigma_1^2), \quad \epsilon_{ij} \sim \mathcal{N}(0, \sigma^2)$$

β_i is the retrieval *system effect*, b_j the *query effect* and ϵ_{ij} the residual error. The system effect is a so-called fixed effects since the systems we would like to compare are fixed and not a sample from some system distribution. On the other hand the query effect is a random effect, assuming that the topic is actually sampled from a population of queries. The assumption behind this model is that the random variable, b_j, is independent and identically normally distributed with zero mean and σ_1^2 variance. The residual error is also assumed to be independent and identically normally distributed with zero mean and σ^2 variance.

Using the t-test to measure the statistical significance of an effect an algorithm has on the search quality, when there are multiple sources of variance, erroneously conflates the variance of the other sources with the variance of the residual error. However, the above-described linear model can be expanded to account for multiple sources of variance:

$$y_{ijk} = \beta_i + b_j + c_{ij} + \epsilon_{ijk}$$

$$b_j \sim \mathcal{N}(0, \sigma_1^2) \quad c_{ij} \sim \mathcal{N}(0, \sigma_1^2) \quad \epsilon_{ijk} \sim \mathcal{N}(0, \sigma^2)$$

c_{ij} now models the within-query effect, given that for a system i and a query j we now have repeated measurements.

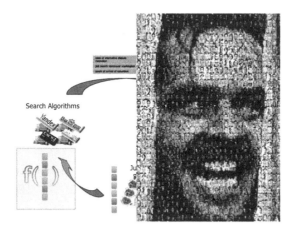

Fig. 15. Point estimates of user model parameters.

The model parameters of evaluation measures are another plausible source of variance. The values of these parameters are either predefined in an ad-hoc way, or point estimates of them are computed by a number of different approaches e.g. by minimise variance in evaluation [87], or fitting a model to gaps between clicks [150]. In all cases however a single value for the parameters is used and hence the evaluation is performed with respect to an average user (Fig. 15). Users however behave very differently when searching, and hence a distribution of these model parameter values could be used instead. In this case measurements do not vary across queries only but for each query they vary due to the different user model parameters used. The t-test linear regression model can be extended again to accommodate for this second source of variance [32, 33].

5   In-situ Evaluation

One of the basic assumptions behind collection-based evaluation is that (a) assessors are actual users of the search engine agree as to what constitutes a relevant and irrelevant document, given a user's query, and (b) the evaluation measures used to quantify the system performance agree with the actual user's satisfaction with the system. In tasks such as search personalisation, or when the search task requires an expertise on the topic this is a drastic assumption to make. In-situ evaluation, on the other hand, removes such an assumption since the user that poses the query is the one deciding about the quality of the returned results. However, obtaining explicit feedback by the user regarding search performance is typically hard, and hence in-situ evaluation depends on the assumption that it is the observable (by the search engine) user behaviour that reflects relevance. Implicit in this is that users behave rationally, that is when searching they have a specific goal in mind and they consistently work towards that goal; they do not behave in randomly or maliciously. This assumption gives high fidelity to in-situ evaluation; users replace judges, evaluation is performed on actual queries typed in, and no ambiguity is present in the posed queries. On the other hand, given that feedback is not explicit one needs to carefully design methods that distill search quality from the noisy implicit user feedback.

There many different user signals that have been used to quantify relevance including:

- Clicks
- Mouse movement
- Browser actions, such as bookmarks, saves, prints
- Dwell time on the SERP and on landing pages of clicked results
- Explicit judgments, such as likes, favourites, etc.
- Query reformulations

5.1   Observable User Behaviour

[42, 50, 60, 67, 68] In this section we will focus on a number of these signals before coming to the experimental design and implementation details of in-situ experiments.

Interpreting Clicks: Clicks on search results is the most exploited user feedback be in-situ evaluation algorithms. The question that arises is what can we infer by observing a user clicking on a search result. How can we assess the quality of the results and how can we translate that to quantifying the quality of the search engine? Let's take Fig. 16(a) as an examples. The user has posed the query "citation metrics" and clicked on the second and fourth result returned by the search engine. Are these clicks an implicit feedback of the perceived by the user relevance of the two results? Are these two clicked results equally good? In a second example, Fig. 16(b), the user posed the query "weather amsterdam" and never clicked on anything. What does a "no click" mean with respect to the quality of the returned results?

(a) Query: "citation metrics"   (b) Query: "weather amsterdam"

Fig. 16. Clicks and no clicks on search engine results pages.

Research focused on interpreting clicks suggest that clicks are noisy; they don't always mean relevance, while their absence is not always a negative signal about the quality of the search algorithm [72]. Further, clicks are biased in a number of ways [85,86]: (a) there is position bias; users are more inclined to examine and click on higher-ranked results, (b) there is contextual bias; whether users click on a result depends on other nearby results, and (c) there is attention bias; users click more on results which draw attention to themselves. However, as we will see later in this section, in the long run clicks point in the right direction.

Interpreting Dwell Time: One of the open questions in interpreting clicks is actually identifying when a click is made on a satisfactory results (SAT click) and when not. Fox et al. [63] tested a large number of implicit feedback to conclude that the two most predictive user signals regarding the satisfaction of the user when clicking a document is the dwell time, that is the time spent on the landing page of the clicked result, and the exit type, i.e. the way in which the user exited the landing page (e.g. by killing the browser window, by typing in a new query, by navigate using history or by session time out). Further a threshold of 30 seconds spent on a landing page appeared to be a good threshold on dwell time towards predicting a SAT from a DSAT click.

Work on the same topic by Kelly and Belkin [93] however suggested that this threshold depends both on the actual user and on the type of the task that led to the search. Kim et al. [96] attempted to address this problem by first segmenting user sessions by a number of features, such as query topic, query type, page topic, and reading level attributes, and then modelled dwell time by a Gamma distribution and through Maximum Likelihood Estimation computed the probability of a dwell time given a SAT click, and the same probability given a DSAT. These probabilities were then used by a classifier to predict whether a click belongs to on the former or the latter category.

Interpreting Mouse Movement: Results in the aforementioned research work dictate that dwell time may not suffice in distinguishing SAT from DSAT clicks on search results. Another line of research examines post-click behaviour, and in particular cursor movements and mouse scrolling as indicators of the searchers satisfaction from the landing pages [69].

In particular, Guo and Agichtein discovered that there is a number of patterns that can dictate whether a user is reading, scanning or skipping big parts of the landing page. Reading indicates relevance while scanning or skipping indicate non-relevance. They observed that horizontal mouse movements, still mouse and mouse at the left half of the screen are good indications of reading, while vertical mouse movements, equal distribution of a still mouse across many areas of the landing page and scrolling indicate scanning or skipping. Scanning the landing page followed by reading can also indicates relevance. Based on these observations they extracted a number of features and built a classifier to predict SAT clicks.

Interpreting Mouse Movement on a SERP: Often times the user examines a SERP but does not click on any search result. In this case the question that arises is whether this is a positive or a negative signal towards the relevance of the SERP. Diriye et al. [58] observed that 41 % of the time that no clicks were recorded was due to dissatisfactory results, but 31 % of the time the results were still satisfying - e.g. the users might have found the answer of her query in the snippets. The percentage of good abandonment of the SERP increased even further when considering search on mobile devises.

Huang et al. [81] used mouse cursor movements on the SERP to predict whether whether a SERP has satisfied a user that did not click on any of the results. Having observed that mouse cursor correlates well with eye gaze, they examined the time spent by a user on hovering over each search result title and the time taken to reach each result title in the ranked list. They discovered that result hover features actually correlate better with human relevance judgments than click-through rates. In addition, even when there are no clicks for a query, hover features show a reasonable correlation with human judgments.

Predicting user satisfaction by the returned SERP requires engineering complex features functions of the position of the mouse, the speed of the mouse, etc. Lagun et al. [102] and Liu et al. [107] proposed the discovery of common mouse subsequences (motifs) and predict satisfaction based on these motifs.

Interpreting Query Reformulations: Hassan et al. [72] examined whether query reformulations can predict user satisfaction by a presented SERP. Specifically, given a query Q_1, a SERP, and a query reformulation Q_2, the goal of their work is to predict the SERP level satisfaction. The considered two classes of features based on (a) the similarity between the two queries, and (b) the time different between the two query submissions. Combining these features with click features they trained a classifier to predict user satisfaction.

In-situ Evaluation Experimental Designs

Clicks, dwell time, mouse movements, reformulations are all signals of user satisfaction with the search results. The question that remains open however is how could one use these signals to compare two search algorithms. There are two predominant frameworks for in-situ evaluation, (a) A/B Testing, and (b) Interleaving.

5.2   A/B Testing

The concept of an A/B testing[6] is trivial. The search traffic is randomly split between two (or more) versions of a search algorithm, the control - production system, and the treatment(s) - experimental system(s). Measures of interest based on observable user behaviour are then collected and analysed. A/B testing is one of the best scientific way to prove causality, i.e., the changes in measures are caused by changes introduced in the treatment(s), and hence it is performed by most commercial search engines[7] to test the quality of their search algorithms. Given that searchers are directed to different versions of the search engine, A/B tests are also used to test the search interface, and other auxiliary search tools (e.g. query recommendation, query autocompletion), and additional panels (e.g. one-box results, knowledge cards) that appear in the SERP.

---

[6] Also known as split testing, control/treatment testing, bucket testing, randomised experiments, and online field experiments.

[7] Amazon, eBay, Etsy, Facebook, Google, Groupon, Intuit, LinkedIn, Microsoft, Net-Flix, Shop Direct, Yahoo!, Zynga have reported performing A/B tests.

Experimental Setup: The simplest setup for an A/B test is to evaluate one factor with two levels, e.g. two different versions of a ranking function. The percentage of traffic directed to the production system versus the experimental system can vary based on (a) the risk on is willing to take by diverting traffic to an experimental version of the search engine, and (b) the statistical power of the experiment towards identifying significant effects - maximum power can be achieved by a 50 %–50 % split. In any case however, the percentage should remain fixed throughout the experiment to avoid observing the Simpsons paradox [115].

The simplest experimental setup for an A/B test is a single parameter experiment. A single parameter experiment means that one is allowed to change a single parameter of the retrieval system, and each change of this parameter generates a different experimental system. Then the search traffic is split between the production and the experimental systems, and hence each user (or, as we will see later, each randomisation unit) is in a single experiment.

Unfortunately, modern retrieval systems have hundreds or even thousands of such parameters; hence this simple single parameter setup simply does not scale. On the other end of complexity lies a multi-factorial design. This is a fully factorial design with N parameter varying across k values per factor. If parameters are independent with each other the that results in N^k experiments, and each randomisation unit is simultaneously in N experiments.

Unfortunately, not all parameters are independent. For instance using a lightweight index that does not store the positions of terms in a document does not allow the efficient use of position-based language model for retrieval. Hence, in reality parameters are partitioned into subsets (layers) of dependent parameters. In this case each randomisation unit is in M independent experiments (for M layers) [16, 56, 98, 139].

Given the experimental infrastructure one actually needs to make a number of decisions when setting up an experiment [51, 97–101]. Some of these decisions are discussed below.

Traffic Diverion: Experiments must specify what subset of traffic is diverted. One easy way to do experiment diversion is random traffic, however this may lead to inconsistent experiences for the users in the case that one query in the session is to the production system while the next query is diverted to the experimental system. A simple mechanism to avoid that is to divert traffic based on the cookies (or login information if available).

Choosing Measure: As mentioned in the previous session measures in in-situ evaluation are using observable user behaviour, such as clicks, mouse movements or query reformulations. Translating these signals into result relevance and using collection-based evaluation measure is one way to measure search quality. A different way that is typically used is to use measures that reflect search quality as a direct function of the user's implicit feedback. These measures can be designed to quantify the overall quality of the SERP (e.g. Click Through Rate, Time to Click, Reciprocal rank of first click, etc.), or the overall experience of the user with the search engine (e.g. number of sessions per user, absence time [38, 138]).

Often times online measures conflict with each other as to whether an experimental system is better than a baseline system, and hence careful analysis of the experimental results is required.

Control Extraneous Factors: There may be many factors that affect the measured quality of a search engine. Test factors are those that are intentionally varied to determine their effects. An example of a non-test factor is the day of the week an experiment is run. During weekend days users typically exhibit different behaviour while search than during week days. Non-test factors can be fixed (e.g. run an experiment only over week days), or integrated out of the experiment (e.g. run an experiment over an entire week). They can be also stratified (run two versions of the experiment, one over week days and one over weekend days, with different intensity, focusing mostly on the stratum in which the experimental system is expected to have higher effect). The latter can increase the statistical power of the experiment.

Estimate Adequate Sample Size: Designing an experiment that enables the discovery of statistical significant effects is of paramount importance. Looking back at section in which statistical significance was discussed, the power of an experiment is a function of the sample size (that is the number of measurements we collect), the effect size one wants to be able to detect (which is the difference of means of the measurements for the two versions of the search engine divided by their standard deviations), and the Type I error, which is typically set to 5%. Pre-setting the power of the experiment (typically set to 80%) and knowing the variance of the measurements can allow us to calculate the number of measurements required. Having specified the number of randomisation units to be diverted on the basis of the risk one wants to endure this defines the length of the experiment.

To estimate the variance one may run an A/A test, that is use the same production system both as a control and as a treatment just to collect data to calculate the variance. Different measures demonstrate different levels of variability, hence the choice of the measure also affects the duration of the experiment. Simple SERP-lecel measures may have smaller variance that overall evaluation measures (such as absence time).

Dilution: There are unavoidable gaps between "showing the feature to users" and "the users experiencing the feature". By focusing on the users that experience a new feature in the case bucket the power and sensitivity of the A/B test increases.

Carry Over Effect: Experiments running in the past may affect users' behaviour in the new experiments. A special case is carry over effects during iterative experimentation, i.e. between different versions of the same experiment with the population in the case buckets dropping off the experiment. To diagnose carry over effects one can test for bucket size abnormality by splitting users to first comers and returning, and test whether the ratio of first comers to returning users remains stable. If abnormality occurs users should be shuffled between experiments and re-run the experiments.

Novelty Impact: Short term user behaviour may not be a good indicator of long term user behaviour: bias can be due to curiosity, learning curve. Hence, it may be the case that even though the power analysis made during the experimental design dictates that an effect should be discovered when running the experiment for a certain period, and even though the effect is actually there, it is not observed. To diagnose a novelty effect the ratio of control/treatment measure throughout experiment is calculated. By considering the second half of the test stable, building confidence intervals one can observe whether values early in the experiment are outside those intervals and continue running the experiment and exclude those early values.

Analysis: Having run the experiment the treatment effect (as a percent change) with 95 % confidence intervals can be used to decide whether any change in the algorithm has resulted in a statistically significant improvement of the perceived by the user search quality.

By construction, base on statistical significance testing, one should expect a 5 % of false positives, that is 5 % of the times the experiment will designate an significant effect which such an effect is not actually there. This 5 % of false positives however assumes that the experiment is run only once, under one dataset, one outcome and one analysis. In A/B testing however this assumption is typically violated: multiple testings are performed, multiple treatments are considered, and multiple measures are reported. Let's assume, for the sake of demonstrating the issue here, that the null hypothesis is actually true, i.e. the experimental system does not improve search quality compared to the production system. The probability of concluding it is false after one test is 0.05. The probability of concluding it is false after two tests is $.05 + .95*.05 = .0975$, while after 90 tests it is 0.99^8. To avoid such erroneous conclusions one needs to adjust the p-values up for to account for the multiple comparisons. There are many different approaches to do that, e.g. the Bonferroni correction, the Tukeys Honest Significant Differences, and the Multivariate t test.

Increasing Sensitivity. One of the open questions in A/B testing is how to increase the sensitivity/power of the experiment. A number of methods have been developed towards this direction [15,55,57,59,70,94].

One way is to reduce the variance of the measurements. Deng et al. [57] suggest two methods to do that: (a) by stratification, and (b) by the use of co-variates. A second way to increase sensitivity is by increasing the sample size. Clearly this is something that one should since the mere point of trying to increase sensitivity is to actually reduce the required sample size. Drutsa et al. [59] instead proposed the use of a pseudo-sample so that the number of measurements increase without actually increasing the length of the experiment. To do that they considered a large number of evaluation measures

---

[8] "If you torture the data enough, it will confess to anything", Ronald Harry Coase.

calculated throughout the experiment, run time series analysis on the and extracted features to help them predict future measurement points. Using these predicted measurements (that is increasing the number of measurements in the experiment) increased the sensitivity of the experiment. Obviously the accuracy of the predictions highly affect the extend to which wrong conclusions are drawn. Kharitonov et al. [94] instead adapted statistical methods for sequential testing (the OBrien and Fleming procedure) that allows stopping the experiment as soon as a statistical significant effect is detected.

Counterfactual A/B Testing Evaluation. Evaluating search engines using A/B testing exposes users to experimental algorithms, running the risk of showing dissatisfactory results. Further, the online deployment of experimental algorithms often requires cumbersome engineering work. Last, even though the multi-layer infrastructure allows multiple experiment to be run in parallel, the experimental framework is often unable to scale to number of parameters values one would like to test over. Therefore, typically an experimental funnel is in place, with collection-based evaluation preceding any in-situ evaluation so that only experimental systems with high potential of improvements over the production system are tested online. However, as mentioned earlier collection-based evaluation has less fidelity compared to in-situ evaluation and sometime the two frameworks disagree regarding the effectiveness of retrieval systems.

Ideally, one would still like to perform in-situ evaluation but offline, using collected query logs as benchmark collections. The issue with doing that is that changes in the SERP produced by the experimental system are often unlikely to exist in the stored query logs, since the production system does not return the same results. Finding ways to use the query log as a benchmark collection that allows the computation of online measures is still an open problem. There are two methods that have appeared in this direction. Grotov et al. [65] trains click models on the basis of historical query logs. Click models can be used to infer the performance of a retrieval system using click signals to infer the relevance of a document given a query. The validity of the experimental results depends on how accurately click models can predict clicks. In a different line of research Li et al. [104–106] assume that for a given query a variety of SERPs are present in historical logs, due to the extensive A/B testing typically performed. If their assumption is valid one can simply compute online measures from historical query logs, correcting for any bias due to the different distribution of queries in the log than the queries received by the live system. When the assumption is not valid fuzzy matching between SERPs (e.g. only matching the top-k documents in the ranked lists) can be used to match the SERP the experimental system is producing and the SERPs in the query log for a given query.

5.3   Interleaving

Interleaving [84] is an alternative method to perform in-situ experiments. Different from A/B testing interleaving does not split the search traffic among systems;

instead results from two (or more) algorithms are simultaneously shown to each user, by being interleaved in a single ranked list. User interactions over the interleaved ranked list are then logged and the preference of users for documents coming from one algorithm over documents coming from an algorithm is inferred.

The advantage of interleaving over A/B testing is that it provides a within-subject design, as opposed to the between-subject design of the A/B testing framework, which can drastically reduce variance, and improve the sensitivity of the experiment.

Devising an interleaving experiment requires (a) coming up with a method to interleave the results of two (or more) search algorithms, (b) deciding how to measure the preference of a user for a document, and (c) assigning this credit back to the participating in the experiment search algorithms. Interleaving algorithms differ from each other in one of these perspectives, with a number of methods developed already:

- Balanced interleaving (Joachims et al. in 2006 [86], Radlinski et al. in 2008 [119])
- Team Draft interleaving (Radlinski et al. in 2008 [119])
- Document constraints interleaving (He et al. in 2009 [74])
- Probabilistic interleaving (Hofmann et al. in 2011 [75])
- Optimised interleaving (Radlinski and Craswell in 2013 [118])
- Vertical aware team draft interleaving (Chuklin et al. in 2014 [46])
- Team draft multileaving (Schuth et al. in 2014 [133])
- Optimised multileaving (Schuth et al. in 2014 [133])
- Probabilistic multileaving (Schuth et al. in 2015 [131])
- Generalised Team Draft Interleaving (Kharitonov et al. in 2015 [95])

In all aforementioned interleaving algorithms clicks are used to infer users preference for one or another participating algorithm, however one can envisage that signals discussed in the Observable User Behaviour section that distill SAT clicks (using dwell time, mouse movement over landing pages, or types of reformulations) could also be used. If documents coming from algorithm A receive more clicks than documents coming from algorithm B, then A wins. Recent work in this direction infers more powerful online measures/statistic for interleaving evaluation [132, 149].

One of the challenges in devising methods to interleave two rankings and assign credit is to avoid possible biases coming from the natural biases user clicks exhibit. Randomisation is typically used to avoid such biases and diagnostics are run to empirically validate the absence of them. Two properties that are typically examined through these diagnostics are: (a) Would random clicking consistently prefer one ranking over another? (b) Would rational clicking consistently prefer one ranking over another equally good one? Balanced interleaving, and Team Draft interleaving fail to exhibit correct behaviour in these corner cases.

A methodological way to develop unbiased interleaving algorithms has been developed by Radlinski and Craswell [118]. The authors model the problem of developing an interleaving algorithm as a constraint optimisation problem. Constraints come from desirable properties in corner cases discussed above, while they optimise for the sensitivity of the derived interleaving method.

Kharitonov et al. [94] have also adapted sequential significance testing methods for the early stopping of interleaving experiments.

Counterfactual Interleaving Evaluation. Similar to the A/B testing framework, one question that arises here is whether interleaving can be performed offline, by using historical query logs as benchmark collections. Hofmann et al. [76] developed a probabilistic interleaving algorithms, similar to Team Draft interleaving, where a coin is tossed at every step of the algorithm to choose the ranking from which the next document will be picked, but within a ranking documents are sampled without replacement according to a predefined distribution over ranks. Having observed user clicks on the interleaved ranking credit is based on all possible assignments of the resulted ranked list (see Fig. 17).

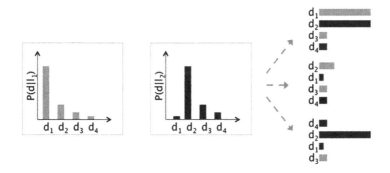

Fig. 17. Possible interleaved rankings in probabilistic interleaving. (Color figure online)

Due to the probabilistic selection of documents to be interleaved, rankings in historical logs may be explained by some permutation of two ranked lists under testing, even though they have not been originated by these algorithms when logged. Importance sampling is used to correct bias. This allows running interleaving experiments using historical logs.

6   Comparative Studies

A valid question to be asked is how these different collection-based and in-situ algorithms compare with each other, and in particular (a) how reliable they are in evaluating the quality of search engines, and (b) how sensitive they are, i.e. how much data do they require to achieve a target confidence level (p-value) [41,117,119].

One of the early comparative studies was performed between interleaved and collection-based experiments [119]. 4–6 pairs of ranking functions, of known retrieval quality, by construction or by judged evaluation, were selected to compare in different settings. Observed user behaviour in two experimental conditions: (a) randomly used one of the two individual ranking functions, and (b) presented an interleaving of the two ranking functions, with the goal to compare all three methods. Evaluation performed on arXiv.org (an academic paper repository) but later on it was replicated against two other search platforms, Bing Web search [117], and Yahoo! Web search [41].

Interleaving Versus Collection-based Evaluation: Experiments on Bing [117] demonstrated that there is a good correlation between expert judgments and interleaving, and further on that one expert judged query is worth approximately 10 queries with clicks.

Interleaving Versus A/B Testing: In the same set of studies interleaving was compared against A/B testing. Comparison were performed on arXiv.org [119], and later replicated on Yahoo! Web Search [41]. Measures computed for the A/B testing included abandonment rate, clicks per query, click at the top-ranked result, pSkip, max and mean Reciprocal Rank of first click, time to first click, and time to last click. Results suggested that the aforementioned measures in A/B testing appeared to be soetimes inconsistent to what was expected (by construction of the rankings under comparison), while results produced by interleaving were at all cases consistent.

Acknowledgements. This work is based on a tutorial I gave at the 2015 Russian Summer School in Information Retrieval (RuSSIR 2015). I would like to thank Ben Carterette, Emine Yilmaz, Anne Schuth, Katja Hofmann, and Filip Radlinski for sharing references and material used in that tutorial and hence as the basis for this survey.

References

1. Agrawal, R., Gollapudi, S., Halverson, A., Ieong, S.: Diversifying search results. In: WSDM, pp. 5–14 (2009)
2. Al-Harbi, A.L., Smucker, M.D.: A qualitative exploration of secondary assessor relevance judging behavior. In: Proceedings of the 5th Information Interaction in Context Symposium, IIiX 2014, pp. 195–204. ACM, New York (2014). http://doi.acm.org/10.1145/2637002.2637025
3. Allan, J., Carterette, B., Dachev, B., Aslam, J.A., Pavlu, V., Kanoulas, E.: Million query track 2007 overview. In: Proceedings of the Sixteenth Text REtrieval Conference, TREC 2007, Gaithersburg, Maryland, USA, 5–9 November 2007. http://trec.nist.gov/pubs/trec16/papers/1MQ.OVERVIEW16.pdf

4. Alonso, O., Baeza-Yates, R.: Design and implementation of relevance assessments using crowdsourcing. In: Clough, P., Foley, C., Gurrin, C., Jones, G.J.F., Kraaij, W., Lee, H., Mudoch, V. (eds.) ECIR 2011. LNCS, vol. 6611, pp. 153–164. Springer, Heidelberg (2011). http://dx.doi.org/10.1007/978-3-642-20161-5_16

5. Alonso, O., Mizzaro, S.: Using crowdsourcing for trec relevance assessment. Inf. Process. Manage. **48**(6), 1053–1066 (2012). http://dx.doi.org/10.1016/j.ipm.2012.01.004

6. Amigó, E., Gonzalo, J., Verdejo, F.: A general evaluation measure for document organization tasks. In: Proceedings of the 36th International ACM SIGIR Conference on Research and Development in Information Retrieval, SIGIR 2013, pp. 643–652. ACM, New York (2013). http://doi.acm.org/10.1145/2484028.2484081

7. Ashkan, A., Clarke, C.L.: On the informativeness of cascade and intent-aware effectiveness measures. In: Proceedings of the 20th International Conference on World Wide Web, WWW 2011, pp. 407–416. ACM, New York (2011). http://doi.acm.org/10.1145/1963405.1963464

8. Aslam, J.A., Pavlu, V., Savell, R.: A unified model for metasearch, pooling, and system evaluation. In: Proceedings of the Twelfth International Conference on Information and Knowledge Management, CIKM 2003, pp. 484–491. ACM, New York (2003). http://doi.acm.org/10.1145/956863.956953

9. Aslam, J.A., Pavlu, V., Yilmaz, E.: A statistical method for system evaluation using incomplete judgments. In: SIGIR 2006: Proceedings of the 29th Annual International ACM SIGIR Conference on Research and Development in Information Retrieval, Seattle, Washington, USA, pp. 541–548, 6–11 August 2006. http://doi.acm.org/10.1145/1148170.1148263

10. Aslam, J.A., Savell, R.: On the effectiveness of evaluating retrieval systems in the absence of relevance judgments. In: Proceedings of the 26th Annual International ACM SIGIR Conference on Research and Development in Informaion Retrieval, SIGIR 2003, pp. 361–362. ACM, New York (2003). http://doi.acm.org/10.1145/860435.860501

11. Aslam, J.A., Yilmaz, E.: Inferring document relevance via average precision. In: Proceedings of the 29th Annual International ACM SIGIR Conference on Research and Development in Information Retrieval, SIGIR 2006, pp. 601–602. ACM, New York (2006). http://doi.acm.org/10.1145/1148170.1148275

12. Aslam, J.A., Yilmaz, E.: Inferring document relevance from incomplete information. In: Proceedings of the Sixteenth ACM Conference on Information and Knowledge Management, CIKM 2007, Lisbon, Portugal, pp. 633–642, 6–10 November 2007. http://doi.acm.org/10.1145/1321440.1321529

13. Aslam, J.A., Yilmaz, E., Pavlu, V.: The maximum entropy method for analyzing retrieval measures. In: Proceedings of the 28th Annual International ACM SIGIR Conference on Research and Development in Information Retrieval, SIGIR 2005, pp. 27–34. ACM, New York (2005). http://doi.acm.org/10.1145/1076034.1076042

14. Bailey, P., Craswell, N., Soboroff, I., Thomas, P., de Vries, A.P., Yilmaz, E.: Relevance assessment: are judges exchangeable and does it matter. In: Proceedings of the 31st Annual International ACM SIGIR Conference on Research and Development in Information Retrieval, SIGIR 2008, Singapore, pp. 667–674, 20–24 July 2008. http://doi.acm.org/10.1145/1390334.1390447

15. Bakshy, E., Eckles, D.: Uncertainty in online experiments with dependent data: an evaluation of bootstrap methods. In: Proceedings of the 19th ACM SIGKDD International Conference on Knowledge Discovery and Data Mining, KDD 2013, pp. 1303–1311. ACM, New York (2013). http://doi.acm.org/10.1145/2487575.2488218

16. Bakshy, E., Eckles, D., Bernstein, M.S.: Designing and deploying online field experiments. In: Proceedings of the 23rd International Conference on World Wide Web, WWW 2014, pp. 283–292. ACM, New York (2014). http://doi.acm.org/10.1145/2566486.2567967

17. Baskaya, F., Keskustalo, H., Järvelin, K.: Simulating simple and fallible relevance feedback. In: Clough, P., Foley, C., Gurrin, C., Jones, G.J.F., Kraaij, W., Lee, H., Mudoch, V. (eds.) ECIR 2011. LNCS, vol. 6611, pp. 593–604. Springer, Heidelberg (2011). http://dl.acm.org/citation.cfm?id=1996889.1996965

18. Baskaya, F., Keskustalo, H., Järvelin, K.: Time drives interaction: simulating sessions in diverse searching environments. In: Proceedings of the 35th International ACM SIGIR Conference on Research and Development in Information Retrieval, SIGIR 2012, pp. 105–114. ACM, New York (2012). http://doi.acm.org/10.1145/2348283.2348301

19. Belkin, N.J.: Salton award lecture: people, interacting with information. In: Proceedings of the 38th International ACM SIGIR Conference on Research and Development in Information Retrieval, Santiago, Chile, pp. 1–2, 9–13 August 2015. http://doi.acm.org/10.1145/2766462.2767854

20. Berto, A., Mizzaro, S., Robertson, S.: On using fewer topics in information retrieval evaluations. In: Proceedings of the 2013 Conference on the Theory of Information Retrieval, ICTIR 2013, pp. 9:30–9:37. ACM, New York (2013). http://doi.acm.org/10.1145/2499178.2499184

21. Bilgic, M., Bennett, P.N.: Active query selection for learning rankers. In: Proceedings of the 35th International ACM SIGIR Conference on Research and Development in Information Retrieval, SIGIR 2012, pp. 1033–1034. ACM, New York (2012). http://doi.acm.org/10.1145/2348283.2348455

22. Blanco, R., Halpin, H., Herzig, D.M., Mika, P., Pound, J., Thompson, H.S., Tran Duc, T.: Repeatable and reliable search system evaluation using crowdsourcing. In: Proceedings of the 34th International ACM SIGIR Conference on Research and Development in Information Retrieval, SIGIR 2011, pp. 923–932. ACM, New York (2011). http://doi.acm.org/10.1145/2009916.2010039

23. Busin, L., Mizzaro, S.: Axiometrics: an axiomatic approach to information retrieval effectiveness metrics. In: Proceedings of the 2013 Conference on the Theory of Information Retrieval, ICTIR 2013, pp. 8:22–8:29. ACM, New York (2013). http://doi.acm.org/10.1145/2499178.2499182

24. Carterette, B.: Robust test collections for retrieval evaluation. In: SIGIR 2007: Proceedings of the 30th Annual International ACM SIGIR Conference on Research and Development in Information Retrieval, Amsterdam, The Netherlands, pp. 55–62, 23–27 July 2007. http://doi.acm.org/10.1145/1277741.1277754

25. Carterette, B.: System effectiveness, user models, and user utility: a conceptual framework for investigation. In: Proceedings of the 34th International ACM SIGIR Conference on Research and Development in Information Retrieval, SIGIR 2011, pp. 903–912. ACM, New York (2011). http://doi.acm.org/10.1145/2009916.2010037

26. Carterette, B.: Multiple testing in statistical analysis of systems-based information retrieval experiments. ACM Trans. Inf. Syst. **30**(1), 4:1–4:34 (2012). http://doi.acm.org/10.1145/2094072.2094076

27. Carterette, B.: Statistical significance testing in information retrieval: theory and practice. In: Proceedings of the 2013 Conference on the Theory of Information Retrieval, ICTIR 2013, p. 2:2. ACM, New York (2013). http://doi.acm.org/10.1145/2499178.2499204

28. Carterette, B.: Statistical significance testing in information retrieval: theory and practice. In: Proceedings of the 37th International ACM SIGIR Conference on Research and Development in Information Retrieval, SIGIR 2014, p. 1286. ACM, New York (2014). http://doi.acm.org/10.1145/2600428.2602292
29. Carterette, B., Allan, J., Sitaraman, R.K.: Minimal test collections for retrieval evaluation. In: SIGIR 2006: Proceedings of the 29th Annual International ACM SIGIR Conference on Research and Development in Information Retrieval, Seattle, Washington, USA, pp. 268–275, 6–11 August 2006. http://doi.acm.org/10.1145/1148170.1148219
30. Carterette, B., Bah, A., Zengin, M.: Dynamic test collections for retrieval evaluation. In: Proceedings of the 2015 International Conference on the Theory of Information Retrieval, ICTIR 2015, pp. 91–100. ACM, New York (2015). http://doi.acm.org/10.1145/2808194.2809470
31. Carterette, B., Kanoulas, E., Pavlu, V., Fang, H.: Reusable test collections through experimental design. In: Proceeding of the 33rd International ACM SIGIR Conference on Research and Development in Information Retrieval, SIGIR 2010, Geneva, Switzerland, pp. 547–554, 19–23 July 2010. http://doi.acm.org/10.1145/1835449.1835541
32. Carterette, B., Kanoulas, E., Yilmaz, E.: Simulating simple user behavior for system effectiveness evaluation. In: Proceedings of the 20th ACM International Conference on Information and Knowledge Management, CIKM 2011, pp. 611–620. ACM, New York (2011). http://doi.acm.org/10.1145/2063576.2063668
33. Carterette, B., Kanoulas, E., Yilmaz, E.: Incorporating variability in user behavior into systems based evaluation. In: Proceedings of the 21st ACM International Conference on Information and Knowledge Management, CIKM 2012, pp. 135–144. ACM, New York (2012). http://doi.acm.org/10.1145/2396761.2396782
34. Carterette, B., Pavlu, V., Fang, H., Kanoulas, E.: Million query track 2009 overview. In: Proceedings of The Eighteenth Text REtrieval Conference, TREC 2009, Gaithersburg, Maryland, USA, 17–20 November 2009. http://trec.nist.gov/pubs/trec18/papers/MQ09OVERVIEW.pdf
35. Carterette, B., Pavlu, V., Kanoulas, E., Aslam, J.A., Allan, J.: Evaluation over thousands of queries. In: Proceedings of the 31st Annual International ACM SIGIR Conference on Research and Development in Information Retrieval, SIGIR 2008, Singapore, pp. 651–658, 20–24 July 2008. http://doi.acm.org/10.1145/1390334.1390445
36. Carterette, B., Pavlu, V., Kanoulas, E., Aslam, J.A., Allan, J.: If I had a million queries. In: Boughanem, M., Berrut, C., Mothe, J., Soule-Dupuy, C. (eds.) ECIR 2009. LNCS, vol. 5478, pp. 288–300. Springer, Heidelberg (2009). http://dx.doi.org/10.1007/978-3-642-00958-7_27
37. Carterette, B., Soboroff, I.: The effect of assessor error on IR system evaluation. In: Proceedings of the 33rd International ACM SIGIR Conference on Research and Development in Information Retrieval, SIGIR 2010, pp. 539–546. ACM, New York (2010). http://doi.acm.org/10.1145/1835449.1835540
38. Chakraborty, S., Radlinski, F., Shokouhi, M., Baecke, P.: On correlation of absence time and search effectiveness. In: Proceedings of the 37th International ACM SIGIR Conference on Research and Development in Information Retrieval, SIGIR 2014, pp. 1163–1166. ACM, New York (2014). http://doi.acm.org/10.1145/2600428.2609535

39. Chandar, P., Webber, W., Carterette, B.: Document features predicting assessor disagreement. In: Proceedings of the 36th International ACM SIGIR Conference on Research and Development in Information Retrieval, SIGIR 2013, pp. 745–748. ACM, New York (2013). http://doi.acm.org/10.1145/2484028.2484161

40. Chapelle, O., Ji, S., Liao, C., Velipasaoglu, E., Lai, L., Wu, S.L.: Intent-based diversification of web search results: metrics and algorithms. Inf. Retr. **14**(6), 572–592 (2011)

41. Chapelle, O., Joachims, T., Radlinski, F., Yue, Y.: Large-scale validation and analysis of interleaved search evaluation. ACM Trans. Inf. Syst. **30**(1), 6:1–6:41 (2012). http://doi.acm.org/10.1145/2094072.2094078

42. Chapelle, O., Metlzer, D., Zhang, Y., Grinspan, P.: Expected reciprocal rank for graded relevance. In: Proceedings of the 18th ACM Conference on Information and Knowledge Management, CIKM 2009, pp. 621–630. ACM, New York (2009). http://doi.acm.org/10.1145/1645953.1646033

43. Chuklin, A., Markov, I., de Rijke, M.: Click Models for Web Search. Synthesis Lectures on Information Concepts, Retrieval, and Services. Morgan & Claypool Publishers, San Rafael (2015). http://dx.doi.org/10.2200/S00654ED1V01Y201507ICR043

44. Chuklin, A., Markov, I., de Rijke, M.: Click Models for Web Search. Synthesis Lectures on Information Concepts, Retrieval, and Services. Morgan & Claypool Publishers, San Rafael (2015). http://clickmodels.weebly.com/uploads/5/2/2/5/52257029/mc2015-clickmodels.pdf

45. Chuklin, A., Serdyukov, P., de Rijke, M.: Click model-based information retrieval metrics. In: Proceedings of the 36th International ACM SIGIR Conference on Research and Development in Information Retrieval, SIGIR 2013, pp. 493–502. ACM, New York (2013). http://doi.acm.org/10.1145/2484028.2484071

46. Chuklin, A., Zhou, K., Schuth, A., Sietsma, F., de Rijke, M.: Evaluating intuitiveness of vertical-aware click models. In: Proceedings of the 37th International ACM SIGIR Conference on Research & Development in Information Retrieval, SIGIR 2014, pp. 1075–1078. ACM, New York (2014). http://doi.acm.org/10.1145/2600428.2609513

47. Clarke, C.L., Kolla, M., Cormack, G.V., Vechtomova, O., Ashkan, A., Büttcher, S., MacKinnon, I.: Novelty and diversity in information retrieval evaluation. In: SIGIR 2008: Proceedings of the 31st annual international ACM SIGIR conference on Research and development in information retrieval, pp. 659–666. ACM, New York (2008)

48. Cormack, G.V., Palmer, C.R., Clarke, C.L.A.: Efficient construction of large test collections. In: Proceedings of the 21st Annual International ACM SIGIR Conference on Research and Development in Information Retrieval, SIGIR 1998, pp. 282–289. ACM, New York (1998). http://doi.acm.org/10.1145/290941.291009

49. Craswell, N., Szummer, M.: Random walks on the click graph. In: Proceedings of the 30th Annual International ACM SIGIR Conference on Research and Development in Information Retrieval, SIGIR 2007, pp. 239–246. ACM, New York (2007). http://doi.acm.org/10.1145/1277741.1277784

50. Craswell, N., Zoeter, O., Taylor, M., Ramsey, B.: An experimental comparison of click position-bias models. In: Proceedings of the 2008 International Conference on Web Search and Data Mining, WSDM 2008, pp. 87–94. ACM, New York (2008). http://doi.acm.org/10.1145/1341531.1341545

51. Crook, T., Frasca, B., Kohavi, R., Longbotham, R.: Seven pitfalls to avoid when running controlled experiments on the web. In: Proceedings of the 15th ACM SIGKDD International Conference on Knowledge Discovery and Data Mining, KDD 2009, pp. 1105–1114. ACM, New York (2009). http://doi.acm.org/10.1145/1557019.1557139

52. Dang, V., Xue, X., Croft, W.B.: Inferring query aspects from reformulations using clustering. In: Proceedings of the 20th ACM International Conference on Information and Knowledge Management, CIKM 2011, pp. 2117–2120. ACM, New York (2011). http://doi.acm.org/10.1145/2063576.2063904

53. Demartini, G., Mizzaro, S.: A classification of IR effectiveness metrics. In: Lalmas, M., MacFarlane, A., Rüger, S.M., Tombros, A., Tsikrika, T., Yavlinsky, A. (eds.) ECIR 2006. LNCS, vol. 3936, pp. 488–491. Springer, Heidelberg (2006). http://dx.doi.org/10.1007/11735106_48

54. Demeester, T., Aly, R., Hiemstra, D., Nguyen, D., Trieschnigg, D., Develder, C.: Exploiting user disagreement for web search evaluation: an experimental approach. In: Proceedings of the 7th ACM International Conference on Web Search and Data Mining, WSDM 2014, pp. 33–42. ACM, New York (2014). http://doi.acm.org/10.1145/2556195.2556268

55. Deng, A., Hu, V.: Diluted treatment effect estimation for trigger analysis in online controlled experiments. In: Proceedings of the Eighth ACM International Conference on Web Search and Data Mining, WSDM 2015, pp. 349–358. ACM, New York (2015). http://doi.acm.org/10.1145/2684822.2685307

56. Deng, A., Li, T., Guo, Y.: Statistical inference in two-stage online controlled experiments with treatment selection and validation. In: Proceedings of the 23rd International Conference on World Wide Web, WWW 2014, pp. 609–618. ACM, New York (2014). http://doi.acm.org/10.1145/2566486.2568028

57. Deng, A., Xu, Y., Kohavi, R., Walker, T.: Improving the sensitivity of online controlled experiments by utilizing pre-experiment data. In: Proceedings of the Sixth ACM International Conference on Web Search and Data Mining, WSDM 2013, pp. 123–132. ACM, New York (2013). http://doi.acm.org/10.1145/2433396.2433413

58. Diriye, A., White, R., Buscher, G., Dumais, S.: Leaving so soon?: understanding and predicting web search abandonment rationales. In: Proceedings of the 21st ACM International Conference on Information and Knowledge Management, CIKM 2012, pp. 1025–1034. ACM, New York (2012). http://doi.acm.org/10.1145/2396761.2398399

59. Drutsa, A., Gusev, G., Serdyukov, P.: Future user engagement prediction and its application to improve the sensitivity of online experiments. In: Proceedings of the 24th International Conference on World Wide Web, WWW 2015, pp. 256–266. International World Wide Web Conferences Steering Committee, Republic and Canton of Geneva (2015). http://dx.doi.org/10.1145/2736277.2741116

60. Dupret, G.E., Piwowarski, B.: A user browsing model to predict search engine click data from past observations. In: Proceedings of the 31st Annual International ACM SIGIR Conference on Research and Development in Information Retrieval, SIGIR 2008, pp. 331–338. ACM, New York (2008). http://doi.acm.org/10.1145/1390334.1390392

61. Efron, M.: Using multiple query aspects to build test collections without human relevance judgments. In: Boughanem, M., Berrut, C., Mothe, J., Soule-Dupuy, C. (eds.) ECIR 2009. LNCS, vol. 5478, pp. 276–287. Springer, Heidelberg (2009). http://dx.doi.org/10.1007/978-3-642-00958-7_26

62. Ferrante, M., Ferro, N., Maistro, M.: Injecting user models and time into precision via markov chains. In: Proceedings of the 37th International ACM SIGIR Conference on Research and Development in Information Retrieval, SIGIR 2014, pp. 597–606. ACM, New York (2014). http://doi.acm.org/10.1145/2600428.2609637

63. Fox, S., Karnawat, K., Mydland, M., Dumais, S., White, T.: Evaluating implicit measures to improve web search. ACM Trans. Inf. Syst. **23**(2), 147–168 (2005). http://doi.acm.org/10.1145/1059981.1059982

64. Grotov, A., Chuklin, A., Markov, I., Stout, L., Xumara, F., de Rijke, M.: A comparative study of click models for web search. In: Mothe, J., Savoy, J., Kamps, J., Pinel-Sauvagnat, K., Jones, G., San Juan, E., Capellato, L., Ferro, N. (eds.) CLEF 2015. LNCS, vol. 9283, pp. 78–90. Springer, Heidelberg (2015). doi:10.1007/978-3-319-24027-5_7

65. Grotov, A., Whiteson, S., de Rijke, M.: Bayesian ranker comparison based on historical user interactions. In: Proceedings of the 38th International ACM SIGIR Conference on Research and Development in Information Retrieval, SIGIR 2015, pp. 273–282. ACM, New York (2015). http://doi.acm.org/10.1145/2766462.2767730

66. Guiver, J., Mizzaro, S., Robertson, S.: A few good topics: Experiments in topic set reduction for retrieval evaluation. ACM Trans. Inf. Syst. **27**(4), 21:1–21:26 (2009). http://doi.acm.org/10.1145/1629096.1629099

67. Guo, F., Liu, C., Kannan, A., Minka, T., Taylor, M., Wang, Y.M., Faloutsos, C.: Click chain model in web search. In: Proceedings of the 18th International Conference on World Wide Web, WWW 2009, pp. 11–20. ACM, New York (2009). http://doi.acm.org/10.1145/1526709.1526712

68. Guo, F., Liu, C., Wang, Y.M.: Efficient multiple-click models in web search. In: Proceedings of the Second ACM International Conference on Web Search and Data Mining, WSDM 2009, pp. 124–131. ACM, New York (2009). http://doi.acm.org/10.1145/1498759.1498818

69. Guo, Q., Agichtein, E.: Beyond dwell time: estimating document relevance from cursor movements and other post-click searcher behavior. In: Proceedings of the 21st International Conference on World Wide Web, WWW 2012, pp. 569–578. ACM, New York (2012). http://doi.acm.org/10.1145/2187836.2187914

70. Guo, Y., Deng, A.: Flexible Online Repeated Measures Experiment. ArXiv e-prints, January 2015

71. Harman, D., Voorhees, E.M.: TREC: an overview. ARIST **40**(1), 113–155 (2006). http://dx.doi.org/10.1002/aris.1440400111

72. Hassan, A., Shi, X., Craswell, N., Ramsey, B.: Beyond clicks: query reformulation as a predictor of search satisfaction. In: Proceedings of the 22nd ACM International Conference on Information and Knowledge Management, CIKM 2013, pp. 2019–2028. ACM, New York (2013). http://doi.acm.org/10.1145/2505515.2505682

73. Hauff, C., Hiemstra, D., Azzopardi, L., de Jong, F.: A case for automatic system evaluation. In: Gurrin, C., He, Y., Kazai, G., Kruschwitz, U., Little, S., Roelleke, T., Rüger, S., van Rijsbergen, K. (eds.) ECIR 2010. LNCS, vol. 5993, pp. 153–165. Springer, Heidelberg (2010). http://dx.doi.org/10.1007/978-3-642-12275-0_16

74. He, J., Zhai, C., Li, X.: Evaluation of methods for relative comparison of retrieval systems based on clickthroughs. In: Proceedings of the 18th ACM Conference on Information and Knowledge Management, CIKM 2009, pp. 2029–2032. ACM, New York (2009). http://doi.acm.org/10.1145/1645953.1646293

75. Hofmann, K., Whiteson, S., de Rijke, M.: A probabilistic method for inferring preferences from clicks. In: Proceedings of the 20th ACM International Conference on Information and Knowledge Management, CIKM 2011, pp. 249–258. ACM, New York (2011). http://doi.acm.org/10.1145/2063576.2063618
76. Hofmann, K., Whiteson, S., de Rijke, M.: Estimating interleaved comparison outcomes from historical click data. In: Proceedings of the 21st ACM International Conference on Information and Knowledge Management, CIKM 2012, pp. 1779–1783. ACM, New York (2012). http://doi.acm.org/10.1145/2396761.2398516
77. Hosseini, M., Cox, I., Milic-Frayling, N.: Optimizing the cost of information retrieval testcollections. In: Proceedings of the 4th Workshop on Workshop for Ph.D. Students in Information and Knowledge Management, PIKM 2011, pp. 79–82. ACM, New York (2011). http://doi.acm.org/10.1145/2065003.2065020
78. Hosseini, M., Cox, I.J., Milic-Frayling, N., Shokouhi, M., Yilmaz, E.: An uncertainty-aware query selection model for evaluation of IR systems. In: Proceedings of the 35th International ACM SIGIR Conference on Research and Development in Information Retrieval, SIGIR 2012, pp. 901–910. ACM, New York (2012). http://doi.acm.org/10.1145/2348283.2348403
79. Hosseini, M., Cox, I.J., Milic-Frayling, N., Vinay, V., Sweeting, T.: Selecting a subset of queries for acquisition of further relevance judgements. In: Amati, G., Crestani, F. (eds.) ICTIR 2011. LNCS, vol. 6931, pp. 113–124. Springer, Heidelberg (2011)
80. Hu, Y., Qian, Y., Li, H., Jiang, D., Pei, J., Zheng, Q.: Mining query subtopics from search log data. In: Proceedings of the 35th International ACM SIGIR Conference on Research and Development in Information Retrieval, SIGIR 2012, pp. 305–314. ACM, New York (2012). http://doi.acm.org/10.1145/2348283.2348327
81. Huang, J., White, R.W., Dumais, S.: No clicks, no problem: using cursor movements to understand and improve search. In: Proceedings of the SIGCHI Conference on Human Factors in Computing Systems, CHI 2011, pp. 1225–1234. ACM, New York (2011). http://doi.acm.org/10.1145/1978942.1979125
82. Järvelin, K., Price, S.L., Delcambre, L.M.L., Nielsen, M.L.: Discounted cumulated gain based evaluation of multiple-query IR sessions. In: Macdonald, C., Ounis, I., Plachouras, V., Ruthven, I., White, R.W. (eds.) ECIR 2008. LNCS, vol. 4956, pp. 4–15. Springer, Heidelberg (2008). http://dl.acm.org/citation.cfm?id=1793274.1793280
83. Jiang, J., He, D., Han, S., Yue, Z., Ni, C.: Contextual evaluation of query reformulations in a search session by user simulation. In: Proceedings of the 21st ACM International Conference on Information and Knowledge Management, CIKM 2012, pp. 2635–2638. ACM, New York (2012). http://doi.acm.org/10.1145/2396761.2398710
84. Joachims, T.: Evaluating retrieval performance using clickthrough data. In: Franke, J., Nakhaeizadeh, G., Renz, I. (eds.) Text Mining, pp. 79–96. Physica/Springer Verlag, New York (2003)
85. Joachims, T., Granka, L., Pan, B., Hembrooke, H., Gay, G.: Accurately interpreting clickthrough data as implicit feedback. In: Proceedings of the 28th Annual International ACM SIGIR Conference on Research and Development in Information Retrieval, SIGIR 2005, pp. 154–161. ACM, New York (2005). http://doi.acm.org/10.1145/1076034.1076063
86. Joachims, T., Granka, L., Pan, B., Hembrooke, H., Radlinski, F., Gay, G.: Evaluating the accuracy of implicit feedback from clicks and query reformulations in web search. ACM Trans. Inf. Syst. 25(2), 1–26 (2007). http://doi.acm.org/10.1145/1229179.1229181

87. Kanoulas, E., Aslam, J.A.: Empirical justification of the gain and discount function for ndcg. In: Proceedings of the 18th ACM Conference on Information and Knowledge Management, CIKM 2009, pp. 611–620. ACM, New York (2009). http://doi.acm.org/10.1145/1645953.1646032

88. Kanoulas, E., Carterette, B., Clough, P.D., Sanderson, M.: Evaluating multi-query sessions. In: Proceedings of the 34th International ACM SIGIR Conference on Research and Development in Information Retrieval, SIGIR 2011, pp. 1053–1062. ACM, New York (2011). http://doi.acm.org/10.1145/2009916.2010056

89. Kazai, G.: In search of quality in crowdsourcing for search engine evaluation. In: Clough, P., Foley, C., Gurrin, C., Jones, G.J.F., Kraaij, W., Lee, H., Mudoch, V. (eds.) ECIR 2011. LNCS, vol. 6611, pp. 165–176. Springer, Heidelberg (2011). http://dl.acm.org/citation.cfm?id=1996889.1996911

90. Kazai, G., Craswell, N., Yilmaz, E., Tahaghoghi, S.: An analysis of systematic judging errors in information retrieval. In: Proceedings of the 21st ACM International Conference on Information and Knowledge Management, CIKM 2012, pp. 105–114. ACM, New York (2012). http://doi.acm.org/10.1145/2396761.2396779

91. Kazai, G., Kamps, J., Milic-Frayling, N.: An analysis of human factors and label accuracy in crowdsourcing relevance judgments. Inf. Retr. **16**(2), 138–178 (2013). http://dx.doi.org/10.1007/s10791-012-9205-0

92. Kazai, G., Yilmaz, E., Craswell, N., Tahaghoghi, S.M.M.: User intent and assessor disagreement in web search evaluation. In: 22nd ACM International Conference on Information and Knowledge Management, CIKM 2013, San Francisco, CA, USA, pp. 699–708, 27 October–1 November 2013. http://doi.acm.org/10.1145/2505515.2505716

93. Kelly, D., Belkin, N.J.: Display time as implicit feedback: understanding task effects. In: Proceedings of the 27th Annual International ACM SIGIR Conference on Research and Development in Information Retrieval, SIGIR 2004, pp. 377–384. ACM, New York (2004). http://doi.acm.org/10.1145/1008992.1009057

94. Kharitonov, V., Macdonald, S., Ounis: sequential testing for early stopping of online experiments. In: Proceedings of the 38th International ACM SIGIR Conference on Research and Development in Information Retrieval, SIGIR 2015. ACM, New York (2015)

95. Kharitonov, E., Macdonald, C., Serdyukov, P., Ounis, I.: Generalized team draft interleaving. In: Proceedings of the 24th ACM International on Conference on Information and Knowledge Management, CIKM 2015, pp. 773–782. ACM, New York (2015). http://doi.acm.org/10.1145/2806416.2806477

96. Kim, Y., Hassan, A., White, R.W., Zitouni, I.: Modeling dwell time to predict click-level satisfaction. In: Proceedings of the 7th ACM International Conference on Web Search and Data Mining, WSDM 2014, pp. 193–202. ACM, New York (2014). http://doi.acm.org/10.1145/2556195.2556220

97. Kohavi, R., Deng, A., Frasca, B., Longbotham, R., Walker, T., Xu, Y.: Trustworthy online controlled experiments: five puzzling outcomes explained. In: Proceedings of the 18th ACM SIGKDD International Conference on Knowledge Discovery and Data Mining, KDD 2012, pp. 786–794. ACM, New York (2012). http://doi.acm.org/10.1145/2339530.2339653

98. Kohavi, R., Deng, A., Frasca, B., Walker, T., Xu, Y., Pohlmann, N.: Online controlled experiments at large scale. In: Proceedings of the 19th ACM SIGKDD International Conference on Knowledge Discovery and Data Mining, KDD 2013, pp. 1168–1176. ACM, New York (2013). http://doi.acm.org/10.1145/2487575.2488217

99. Kohavi, R., Deng, A., Longbotham, R., Xu, Y.: Seven rules of thumb for web site experimenters. In: Proceedings of the 20th ACM SIGKDD International Conference on Knowledge Discovery and Data Mining, KDD 2014, pp. 1857–1866. ACM, New York (2014). http://doi.acm.org/10.1145/2623330.2623341

100. Kohavi, R., Longbotham, R.: Online controlled experiments and A/B tests. In: Sammut, C., Webb, G. (eds.) Encyclopedia of Machine Learning and Data Mining (2015)

101. Kohavi, R., Longbotham, R., Sommerfield, D., Henne, R.: Controlled experiments on the web: survey and practical guide. Data Min. Knowl. Disc. 18(1), 140–181 (2009). http://dx.doi.org/10.1007/s10618-008-0114-1

102. Lagun, D., Ageev, M., Guo, Q., Agichtein, E.: Discovering common motifs in cursor movement data for improving web search. In: Proceedings of the 7th ACM International Conference on Web Search and Data Mining, WSDM 2014, pp. 183–192. ACM, New York (2014). http://doi.acm.org/10.1145/2556195.2556265

103. Lease, M., Yilmaz, E.: Crowdsourcing for information retrieval. SIGIR Forum 45(2), 66–75 (2012). http://doi.acm.org/10.1145/2093346.2093356

104. Li, L., Chen, S., Kleban, J., Gupta, A.: Counterfactual estimation and optimization of click metrics for search engines. CoRR abs/1403.1891 (2014). http://arxiv.org/abs/1403.1891

105. Li, L., Chen, S., Kleban, J., Gupta, A.: Counterfactual estimation and optimization of click metrics in search engines: a case study. In: Proceedings of the 24th International Conference on World Wide Web Companion, WWW 2015 Companion, pp. 929–934. International World Wide Web Conferences Steering Committee, Republic and Canton of Geneva (2015). http://dx.doi.org/10.1145/2740908.2742562

106. Li, L., Kim, J.Y., Zitouni, I.: Toward predicting the outcome of an a/b experiment for search relevance. In: Proceedings of the Eighth ACM International Conference on Web Search and Data Mining, WSDM 2015, pp. 37–46. ACM, New York (2015). http://doi.acm.org/10.1145/2684822.2685311

107. Liu, Y., Chen, Y., Tang, J., Sun, J., Zhang, M., Ma, S., Zhu, X.: Different users, different opinions: predicting search satisfaction with mouse movement information. In: Proceedings of the 38th International ACM SIGIR Conference on Research and Development in Information Retrieval, SIGIR 2015, pp. 493–502. ACM, New York (2015). http://doi.acm.org/10.1145/2766462.2767721

108. Maddalena, E., Mizzaro, S., Scholer, F., Turpin, A.: Judging relevance using magnitude estimation. In: Hanbury, A., Kazai, G., Rauber, A., Fuhr, N. (eds.) ECIR 2015. LNCS, vol. 9022, pp. 215–220. Springer, Heidelberg (2015). http://dx.doi.org/10.1007/978-3-319-16354-3_23

109. Megorskaya, O., Kukushkin, V., Serdyukov, P.: On the relation between assessor's agreement and accuracy in gamified relevance assessment. In: Proceedings of the 38th International ACM SIGIR Conference on Research and Development in Information Retrieval, SIGIR 2015, pp. 605–614. ACM, New York (2015). http://doi.acm.org/10.1145/2766462.2767727

110. Mehrotra, R., Yilmaz, E.: Representative & informative query selection for learning to rank using submodular functions. In: Proceedings of the 38th International ACM SIGIR Conference on Research and Development in Information Retrieval, SIGIR 2015, pp. 545–554. ACM, New York (2015). http://doi.acm.org/10.1145/2766462.2767753

111. Metrikov, P., Pavlu, V., Aslam, J.A.: Impact of assessor disagreement on ranking performance. In: Proceedings of the 35th International ACM SIGIR Conference on Research and Development in Information Retrieval, SIGIR 2012, pp. 1091–1092. ACM, New York (2012). http://doi.acm.org/10.1145/2348283.2348484

112. Moffat, A., Zobel, J.: Rank-biased precision for measurement of retrieval effectiveness. ACM Trans. Inf. Syst. **27**(1), 2:1–2:27 (2008). http://doi.acm.org/10.1145/1416950.1416952

113. Nuray, R., Can, F.: Automatic ranking of information retrieval systems using data fusion. Inf. Process. Manage. **42**(3), 595–614 (2006). http://dx.doi.org/10.1016/j.ipm.2005.03.023

114. Pavlu, V., Rajput, S., Golbus, P.B., Aslam, J.A.: IR system evaluation using nugget-based test collections. In: Proceedings of the Fifth ACM International Conference on Web Search and Data Mining, WSDM 2012, pp. 393–402. ACM, New York (2012). http://doi.acm.org/10.1145/2124295.2124343

115. Pearl, J.: Comment: understanding simpson's paradox. Am. Stat. **68**(1), 8–13 (2014). http://EconPapers.repec.org/RePEc:taf:amstat:v:68:y:2014:i:1:p:8–13

116. Qian, Y., Sakai, T., Ye, J., Zheng, Q., Li, C.: Dynamic query intent mining from a search log stream. In: Proceedings of the 22nd ACM International Conference on Information and Knowledge Management, CIKM 2013, pp. 1205–1208. ACM, New York (2013). http://doi.acm.org/10.1145/2505515.2507856

117. Radlinski, F., Craswell, N.: Comparing the sensitivity of information retrieval metrics. In: Proceedings of the 33rd International ACM SIGIR Conference on Research and Development in Information Retrieval, SIGIR 2010, pp. 667–674. ACM, New York (2010). http://doi.acm.org/10.1145/1835449.1835560

118. Radlinski, F., Craswell, N.: Optimized interleaving for online retrieval evaluation. In: Proceedings of the Sixth ACM International Conference on Web Search and Data Mining, WSDM 2013, pp. 245–254. ACM, New York (2013). http://doi.acm.org/10.1145/2433396.2433429

119. Radlinski, F., Kurup, M., Joachims, T.: How does clickthrough data reflect retrieval quality? In: Proceedings of the 17th ACM Conference on Information and Knowledge Management, CIKM 2008, pp. 43–52. ACM, New York (2008). http://doi.acm.org/10.1145/1458082.1458092

120. Radlinski, F., Szummer, M., Craswell, N.: Inferring query intent from reformulations and clicks. In: Proceedings of the 19th International Conference on World Wide Web, WWW 2010, pp. 1171–1172. ACM, New York (2010). http://doi.acm.org/10.1145/1772690.1772859

121. Robertson, S.: On the contributions of topics to system evaluation. In: Clough, P., Foley, C., Gurrin, C., Jones, G.J.F., Kraaij, W., Lee, H., Mudoch, V. (eds.) ECIR 2011. LNCS, vol. 6611, pp. 129–140. Springer, Heidelberg (2011). http://dl.acm.org/citation.cfm?id=1996889.1996908

122. Robertson, S.E., Kanoulas, E.: On per-topic variance in IR evaluation. In: Proceedings of the 35th International ACM SIGIR Conference on Research and Development in Information Retrieval, SIGIR 2012, pp. 891–900. ACM, New York (2012). http://doi.acm.org/10.1145/2348283.2348402

123. Sakai, T.: Bootstrap-based comparisons of IR metrics for finding one relevant document. In: Ng, H.T., Leong, M.-K., Kan, M.-Y., Ji, D. (eds.) AIRS 2006. LNCS, vol. 4182, pp. 374–389. Springer, Heidelberg (2006). http://dx.doi.org/10.1007/11880592_29

124. Sakai, T.: Designing test collections for comparing many systems. In: Proceedings of the 23rd ACM International Conference on Conference on Information and Knowledge Management, CIKM 2014, pp. 61–70. ACM, New York (2014). http://doi.acm.org/10.1145/2661829.2661893

125. Sakai, T., Dou, Z., Clarke, C.L.: The impact of intent selection on diversified search evaluation. In: Proceedings of the 36th International ACM SIGIR Conference on Research and Development in Information Retrieval, SIGIR 2013, pp. 921–924. ACM, New York (2013). http://doi.acm.org/10.1145/2484028.2484105

126. Sakai, T., Song, R.: Evaluating diversified search results using per-intent graded relevance. In: Proceedings of the 34th International ACM SIGIR Conference on Research and Development in Information Retrieval, SIGIR 2011, pp. 1043–1052. ACM, New York (2011). http://doi.acm.org/10.1145/2009916.2010055

127. Sanderson, M.: Test collection based evaluation of information retrieval systems. Found. Trends Inf. Retrieval **4**(4), 247–375 (2010). http://dx.doi.org/10.1561/1500000009

128. Sanderson, M., Paramita, M.L., Clough, P., Kanoulas, E.: Do user preferences and evaluation measures line up? In: Proceedings of the 33rd International ACM SIGIR Conference on Research and Development in Information Retrieval, SIGIR 2010, pp. 555–562. ACM, New York (2010). http://doi.acm.org/10.1145/1835449.1835542

129. Schaer, P.: Better than their reputation? On the reliability of relevance assessments with students. In: Catarci, T., Forner, P., Hiemstra, D., Peñas, A., Santucci, G. (eds.) CLEF 2012. LNCS, vol. 7488, pp. 124–135. Springer, Heidelberg (2012). http://dx.doi.org/10.1007/978-3-642-33247-0_14

130. Scholer, F., Turpin, A., Sanderson, M.: Quantifying test collection quality based on the consistency of relevance judgements. In: Proceedings of the 34th International ACM SIGIR Conference on Research and Development in Information Retrieval, SIGIR 2011, pp. 1063–1072. ACM, New York (2011). http://doi.acm.org/10.1145/2009916.2010057

131. Schuth, A., Bruintjes, R.J., Buüttner, F., van Doorn, J., Groenland, C., Oosterhuis, H., Tran, C.N., Veeling, B., van der Velde, J., Wechsler, R., Woudenberg, D., de Rijke, M.: Probabilistic multileave for online retrieval evaluation. In: Proceedings of the 38th International ACM SIGIR Conference on Research and Development in Information Retrieval, SIGIR 2015, pp. 955–958. ACM, New York (2015). http://doi.acm.org/10.1145/2766462.2767838

132. Schuth, A., Hofmann, K., Radlinski, F.: Predicting search satisfaction metrics with interleaved comparisons. In: Proceedings of the 38th International ACM SIGIR Conference on Research and Development in Information Retrieval, SIGIR 2015, pp. 463–472. ACM, New York (2015). http://doi.acm.org/10.1145/2766462.2767695

133. Schuth, A., Sietsma, F., Whiteson, S., Lefortier, D., de Rijke, M.: Multileaved comparisons for fast online evaluation. In: Proceedings of the 23rd ACM International Conference on Conference on Information and Knowledge Management, CIKM 2014, pp. 71–80. ACM, New York (2014). http://doi.acm.org/10.1145/2661829.2661952

134. Smucker, M.D., Allan, J., Carterette, B.: A comparison of statistical significance tests for information retrieval evaluation. In: Proceedings of the Sixteenth ACM Conference on Conference on Information and Knowledge Management, CIKM 2007, pp. 623–632. ACM, New York (2007). http://doi.acm.org/10.1145/1321440.1321528

135. Smucker, M.D., Allan, J., Carterette, B.: Agreement among statistical significance tests for information retrieval evaluation at varying sample sizes. In: Proceedings of the 32nd International ACM SIGIR Conference on Research and Development in Information Retrieval, SIGIR 2009, pp. 630–631. ACM, New York (2009). http://doi.acm.org/10.1145/1571941.1572050

136. Smucker, M.D., Clarke, C.L.: Time-based calibration of effectiveness measures. In: Proceedings of the 35th International ACM SIGIR Conference on Research and Development in Information Retrieval, SIGIR 2012, pp. 95–104. ACM, New York (2012). http://doi.acm.org/10.1145/2348283.2348300

137. Soboroff, I., Nicholas, C., Cahan, P.: Ranking retrieval systems without relevance judgments. In: Proceedings of the 24th Annual International ACM SIGIR Conference on Research and Development in Information Retrieval, SIGIR 2001, pp. 66–73. ACM, New York (2001). http://doi.acm.org/10.1145/383952.383961

138. Song, Y., Shi, X., Fu, X.: Evaluating and predicting user engagement change with degraded search relevance. In: Proceedings of the 22Nd International Conference on World Wide Web, WWW 2013, pp. 1213–1224. International World Wide Web Conferences Steering Committee, Republic and Canton of Geneva (2013). http://dl.acm.org/citation.cfm?id=2488388.2488494

139. Tang, D., Agarwal, A., O'Brien, D., Meyer, M.: Overlapping experiment infrastructure: more, better, faster experimentation. In: Proceedings of the 16th ACM SIGKDD International Conference on Knowledge Discovery and Data Mining, KDD 2010, pp. 17–26. ACM, New York (2010). http://doi.acm.org/10.1145/1835804.1835810

140. Turpin, A., Scholer, F., Mizzaro, S., Maddalena, E.: The benefits of magnitude estimation relevance assessments for information retrieval evaluation. In: Proceedings of the 38th International ACM SIGIR Conference on Research and Development in Information Retrieval, SIGIR 2015, pp. 565–574. ACM, New York (2015). http://doi.acm.org/10.1145/2766462.2767760

141. Webber, W., Chandar, P., Carterette, B.: Alternative assessor disagreement and retrieval depth. In: Proceedings of the 21st ACM International Conference on Information and Knowledge Management, CIKM 2012, pp. 125–134. ACM, New York (2012). http://doi.acm.org/10.1145/2396761.2396781

142. Wu, S., Crestani, F.: Methods for ranking information retrieval systems without relevance judgments. In: Proceedings of the 2003 ACM Symposium on Applied Computing, SAC 2003, pp. 811–816. ACM, New York (2003). http://doi.acm.org/10.1145/952532.952693

143. Yilmaz, E., Aslam, J.A.: Estimating average precision with incomplete and imperfect judgments. In: Proceedings of the 2006 ACM CIKM International Conference on Information and Knowledge Management, Arlington, Virginia, USA, pp. 102–111, 6–11 November 2006. http://doi.acm.org/10.1145/1183614.1183633

144. Yilmaz, E., Kanoulas, E., Aslam, J.A.: A simple and efficient sampling method for estimating AP and NDCG. In: Proceedings of the 31st Annual International ACM SIGIR Conference on Research and Development in Information Retrieval, SIGIR 2008, Singapore, pp. 603–610, 20–24 July 2008. http://doi.acm.org/10.1145/1390334.1390437

145. Yilmaz, E., Kanoulas, E., Craswell, N.: Effect of intent descriptions on retrieval evaluation. In: Proceedings of the 23rd ACM International Conference on Conference on Information and Knowledge Management, CIKM 2014, pp. 599–608. ACM, New York (2014). http://doi.acm.org/10.1145/2661829.2661950

146. Yilmaz, E., Kazai, G., Craswell, N., Tahaghoghi, S.M.: On judgments obtained from a commercial search engine. In: Proceedings of the 35th International ACM SIGIR Conference on Research and Development in Information Retrieval, SIGIR 2012, pp. 1115–1116. ACM, New York (2012). http://doi.acm.org/10.1145/2348283.2348496

147. Yilmaz, E., Shokouhi, M., Craswell, N., Robertson, S.: Expected browsing utility for web search evaluation. In: Proceedings of the 19th ACM International Conference on Information and Knowledge Management, CIKM 2010, pp. 1561–1564. ACM, New York (2010). http://doi.acm.org/10.1145/1871437.1871672

148. Yilmaz, E., Verma, M., Craswell, N., Radlinski, F., Bailey, P.: Relevance and effort: an analysis of document utility. In: Proceedings of the 23rd ACM International Conference on Conference on Information and Knowledge Management, CIKM 2014, pp. 91–100. ACM, New York (2014). http://doi.acm.org/10.1145/2661829.2661953

149. Yue, Y., Gao, Y., Chapelle, O., Zhang, Y., Joachims, T.: Learning more powerful test statistics for click-based retrieval evaluation. In: Proceedings of the 33rd International ACM SIGIR Conference on Research and Development in Information Retrieval, SIGIR 2010, pp. 507–514. ACM, New York (2010). http://doi.acm.org/10.1145/1835449.1835534

150. Zhang, Y., Park, L.A., Moffat, A.: Click-based evidence for decaying weight distributions in search effectiveness metrics. Inf. Retr. **13**(1), 46–69 (2010). http://dx.doi.org/10.1007/s10791-009-9099-7

151. Zhu, J., Wang, J., Vinay, V., Cox, I.J.: Topic (query) selection for IR evaluation. In: Proceedings of the 32nd International ACM SIGIR Conference on Research and Development in Information Retrieval, SIGIR 2009, pp. 802–803. ACM, New York (2009). http://doi.acm.org/10.1145/1571941.1572136

Data Science for Massive Networks

Anton Kocheturov[1,2(✉)] and Panos M. Pardalos[1,2]

[1] Center for Applied Optimization, University of Florida,
401 Weil Hall, P.O. Box 116595, Gainesville, FL 32611-6595, USA
`antrubler@gmail.com`
[2] Laboratory of Algorithms and Technologies for Network Analysis,
National Research University Higher School of Economics,
136 Rodionova, Nizhny Novgorod 603093, Russian Federation

Abstract. In this chapter we attempt to briefly describe a history of massive networks, their place in modern life, and discuss open problems related to them. We start with giving a historical overview indicating the most influential milestones in the development of networks. Then we consider how real-life massive datasets can be represented in terms of networks describing some examples and summarizing properties of such networks. We also discuss cases of modeling real-life massive networks. In addition, we give some examples of how to optimize in massive networks and in which areas we can apply these techniques. We conclude by discussing open problems of massive networks.

Keywords: Massive data sets · Networks · Small-world · Power-law · Human brain networks · Market graphs · Call graph · Robustness · Optimization · Cliques · Independent sets · Minimum spanning trees

1 Network Analysis History[1]

Network analysis originated many years ago when Leonhard Euler solved the famous problem of Seven Bridges of Königsberg in 1736. Euler's solution is considered to be the first theorem in the field of network analysis and graph theory. Then in the 19th century Gustav Kirchhoff initiated the theory of electrical networks and defined the flow conservation equations that are one of the milestones of network flow theory today. After the invention of the telephone by Alexander Bell in the end of the same century the resulting applications gave the network analysis a great stimulus.

The field evolved dramatically after the 19th century. As in many other fields, World War II played a crucial role in the development of it as many algorithms and techniques were developed to solve logistics problems from the military. After the war, these technological advances were successfully applied in many other areas.

The earliest linear programming model was developed by Leonid Kantorovich in 1939 during World War II, to plan expenditures to reduce the costs of the army.

---

[1] See Floudas, C.A., Pardalos, P.M.: Encyclopedia of Optimization. Springer (2008).

© Springer International Publishing Switzerland 2016
P. Braslavski et al. (Eds.): RuSSIR 2015, CCIS 573, pp. 88–100, 2016.
DOI: 10.1007/978-3-319-41718-9_4

In 1940, also during World War II, Tjalling Koopmans formulated linear optimization models to select shipping routes to send commodities from America, to Specific destinations in England. For their work in the theory of optimum allocation of resources, these two researchers were awarded with the Nobel price in Economics in 1975.

The first complete algorithm for solving the transportation problem was proposed by Frank Lauren Hitchcock in 1941. With the development of the Simplex Method for solving linear programs by George B. Dantzig in 1957, a new framework for solving several network problems became available. The network simplex algorithm is still in practice and one of the most efficient algorithms for solving network flow problems.

Many other authors proposed efficient algorithms for solving different network problems. Joseph Kruskal in 1956 and Robert C. Prim in 1957 developed algorithms for solving the minimum spanning tree problem. In 1956 Edsger W. Dijkstra developed his algorithm for solving the shortest path problem, possibly one of the most recognized algorithms in network analysis.

Then, as it happened in other fields, computer science and networks influenced each other in many aspects. In 1963 the book by Lester R. Ford and Delbert R. Fulkerson introduced new developments in data structure techniques and computational complexity into the field of networks.

In recent years the evolution of computers have changed the field. We are now able to solve large-scale network problems with the advances in parallel, grid, and cloud computing. The direction of quantum computing is also worth mentioning despite of the fact that a lot of theoretical and engineering work still has to be done before getting the desired speed up.

Network Analysis has become a major research topic over the last years. The broad range of applications that can be described and analyzed by means of a network is bringing together researches from numerous fields: operations research, computer science, transportation, biomedicine, energy, social sciences, computational neuroscience, and others. This remarkable diversity of the fields that take advantage of network analysis makes the endeavor of gathering up-to-date material a very useful task.

2   Graph Representation of Massive Datasets

In many cases it is convenient to represent a dataset as a graph (network) with certain attributes associated with its vertices and edges. Studying the properties of these graphs often provides useful information about the internal structure of the datasets they represent.

In this section we consider some remarkable examples and classes of networks.

2.1   Small-World and Power-Law Phenomena

The empirical evidence of past years showed that the majority of real-life networks possess common properties.

First, they are **small-world networks** [13], saying that they have three following characteristics:

1. **Sparseness.** These networks tend to be sparse, i.e., the density of a network, defined as a ratio of the number of edges in the graph to the maximally possible number of edges in this graph, is close to zero.
2. **High clustering coefficient.** Despite their spareness, they tend to contain cliques, or subgraphs in which every vertex is connected with others. This tendency can be quantified by a clustering coefficient. There are many ways how to define a clustering coefficient [20] but usually the global clustering coefficient C for unweighted undirected graph with no loops and multi-edges is assumed, that is defined as in [36]:

$$C = \frac{\text{\# of closed triplets}}{\text{\# of connected triplets}}, \tag{1}$$

where by a closed (connected) triplet we mean any subgraph consisting of three vertices and three (at least two) edges.

An alternative definition of C that is also often used in the literature can be found in [46].

3. **Short average path length.** Possibly the most important property out of these three. It states that the average shortest path length L is small, where L defined as in [46]:

$$L = \frac{1}{N(N-1)} \sum_{i,j \in E} d_{ij}, \tag{2}$$

where E is the set of vertices of a network consisting of N elements, and d_{ij} is the shortest path between vertices i and j (smallest number of edges one needs to travel to get from vertex i to vertex j). L is usually assumed to grow at most as a logarithm of the network size [46].

Second, they are **power-law** networks [13,14], saying that degrees of vertices of those networks are distributed according to the power-law distribution. In other words, the probability for a vertex to have a degree equal k is:

$$P(k) \propto k^{-\gamma} \tag{3}$$

As a possible consequence of the scale-free and small-world properties, these networks usually have other common characteristics such as tendency to contain hubs (because of the power-law and small diameter), and presence of a giant connected component (because of the power-law [3]). Here by a connected component we mean a subgraph of the initial graph for which there is a path for all possible pairs of vertices. The second notion doesn't have a precise definition but a hub can be think of as a node of a high degree which serves as a connector for a number of other vertices.

2.2   World Wide Web

The World Wide Web is an enormously large dynamic directed graph where vetrices are documents and arcs are links pointing from one vertex to another. The real size of it is an open question but some estimates can be found: for instance, in 2008 it was reported to consist of 10^{12} unique URLs [24]. The fact that this is a directed graph allows us consider in-degree and out-degree distributions separately. Surprisingly, both distributions are reported [5,9,10,22,31] to follow the power-law with parameter $\gamma \in (2,3)$. Another surprising fact is that the parameter values are not equal to each other and the one related to the in-degree (equal 2.1) is substantially smaller than the one related to the out-degree (2.3–2.7).

Another amazing observation was that the in- and out-degree distributions seemed to be independent of the size of the analyzed portion of the web. In other words, the web is a **scale-free** graph which is another important property of the real-life networks.

2.3   Social Networks

Analysis of social networks is useful in many fields. These networks are rich with information since they contain a number of parameters describing people as well as many different types of relations between them.

The first and most natural example to come in one's mind is the "acquaintanceship" graph that would contain a vertex associated with each person on the planet, with an edge connecting two vertices if the corresponding people know each other. The famous "small-world" hypothesis associated with this graph claims that despite the large number of vertices, the maximum distance between any two of them is small. More specifically, the idea of "six degrees of separation" states that any two people in the world are linked with each other through a sequence of at most six people [25,46]. Historically introduced first by Milgram in 1960s [34] (however, it wasn't stated in this way), it is believed to be the first time when the small diameter property was formulated.

While one cannot verify this hypothesis for the acquaintanceship graph, certain subgraphs of this graph can be investigated in details. One example subgraph is the scientific collaboration graph reflecting information about the joint works between all scientists. Two vertices are connected by an edge in this graph if the corresponding two scientists have a joint research paper. A well-known concept associated with this graph is the so-called "Erdös number", which is assigned to every vertex and characterize the minimum distance from this vertex to the "center vertex" of the graph (represented by the famous graph theoretician Paul Erdös)

2.4   Small World Networks in Neuroscience

A single human brain is a complex dynamic network of a tremendous size. It contains a number of different components and connections between them.

The human cerebral cortex was estimated to contain roughly 8×10^9 neurons and 7×10^{13} connections [35]. Moreover, these building elements of a brain are assembled into systems of higher complexity creating the hierarchical multi-layer structure. Despite a number of attempts, analyzing the whole brain is still a backbreaking problem.

However, there are reasonable ways ho to attack this problem. For instance, one can apply network modeling approach to study such a complex system by partitioning the brain into non-overlapping functional brain regions having statistically dependent neural activity patterns [39].

Even analysis of the overall behavior of the brain can be a very fruitful approach in some applications: for example, the chaos level of the brain can be measured to predict epilepsy seizures [26,27,48].

Surprisingly, many networks arising in the area of neuroscience were reported to be small-world scale-free networks [29], that is believed to be due to the fact that brain could have evolutionarily become a robust, efficient from the information point of view, system.

2.5 Call Graph

In a call graph, the vertices are telephone numbers, and two vertices are connected by an edge if a call was made from one number to another. A call graph was constructed with data from AT&T telephone billing records [1,15]. Based on one 20-day period it had 290 million vertices and 4 billion edges. The analyzed one-day call graph had 53,767,087 vertices and over 170 millions of edges.

This graph appeared to have 3,667,448 connected components, most of them tiny. A giant connected component with 44,989,297 vertices (more than 80 % of the total) was computed. The distribution of the degrees of the vertices follows the power-law distribution.

2.6 Market Graphs

Assume that we are to study a financial market of a particular country or the global financial market in general. We can utilize the power of the network approach by constructing a network (graph) related to the market in the following manner: each stock (for a moment we are interested only in stocks, but the approach can be generalized for any financial agent or instrument) corresponds to a vertex in the network. Each link between two stocks (vertices) is represented by a weighted edge, where the weight reflects degree of similarity (can be defined in different ways [12,14]) between stocks. Notice, that in this setting any market network is a complete weighted graph.

Like other real-life networks market graphs share their properties: for example, a market graph of the US stock market was reported to fallow the power-law [14].

3   Modeling Massive Networks

Due to the enormous size of real-life networks, even simple tasks can be very difficult from the computational point of view. Thus modeling such graphs is a critical tool in studying and predicting their properties.

It turned out that the probabilistic approach can be applied successfully for modeling some most important properties of the real-life networks. One may argue that this is due to a stochastic nature of agents (vertices) and interactions between them (arcs or edges) who dynamically form such graphs.

The first way is to consider a family of random graphs having a predefined set of parameters such as, for example, fixed degree distribution and clustering coefficient, and generate a graph out of this family [3,8,17,18,36,38,44]. Two of such models are $G(n, m)$ and $\mathcal{G}(n, p)$ [17,18]. The first model assigns the same probability to all graphs with n vertices and m edges, while in the second model each pair of vertices is chosen to be linked by an edge randomly and independently with probability p.

It it said that the property Q holds asymptotically almost surely (a.a.s.) if the probability that a random graph on n vertices has this property tends to 1 as $n \to \infty$. The asymptotic properties of uniform random graphs have been well studied. For instance, a uniform random graph $\mathcal{G}(n, p)$ is a.a.s. connected if $p > \frac{\log n}{n}$ and it is a.a.s. disconnected otherwise. Also, if p is in the range $\frac{1}{n} < p < \frac{\log n}{n}$, the graph $\mathcal{G}(n, p)$ a.a.s. has a unique giant connected component (i.e., a component having a fixed percentage of the number of nodes n) which is also a very important property of real-life networks. While these uniform random graph models capture some properties of real-world massive graphs (such as the emergence of a giant connected component), they fail to capture other important properties such as high clustering coefficients or the proper distribution of vertices' degree.

Another way is to consider a graph as a result of a stochastic process under some suitable conditions which guarantee the existence of properties of interest. Probably, the most famous and well-studied is the class of the preferential attachment models [4,9,19,30,32,37] (For others, see [6]). The main idea of this approach introduced in [9] is that we construct a graph from the scratch and on each step we add one vertex and connect it with one of the already existing vertices with the probabilities proportional to the degrees of these vertices. In the paper [37] the authors generalized the concept having introduced the so-called *PA*-class (stands for "preferential attachment") and showed that all other models are just cases of it. They also found a particular representative of the class which they called the "Polynomial Model" and for which both global and local clustering coefficients are constant and simultaneously one can tune the parameter γ of the degree distribution to be any from $(2; \infty)$.

All existing models are not restricted to the classes described above: for a comprehensive survey on the topic refer, for example, to [6].

4 Optimization on Very Large Graphs with Applications

4.1 Cliques

Given graph $G = (V, E)$, a subset S of its vertices is called a clique if the subgraph $G(S)$ induced by S is complete; i.e. there is an edge between any two vertices in $G(S)$. For any graph, we are usually interested in two notions related to cliques: a maximal clique that is a clique which is not a proper subset of another clique of this graph; and a maximum clique that is a clique of the maximum cardinality in this graph.

The maximum clique problem (MCP), that is to find a maximum clique in a given graph, is one of the classical problems in graph theory with many applications in different fields including project selection, classification, fault tolerance, coding, computer vision, economics, information retrieval, signal transmission, and alignment of DNA with protein sequences [21]. One of the ways to solve it is to use an Integer Programming formulation as follows:

$$
\begin{aligned}
\max \ & \sum x_i \\
s.t. : \ & x_i + x_j \ \forall (i, j) \notin E \\
& x_i \in \{0; 1\}
\end{aligned}
\tag{4}
$$

Despite the simplicity of the formulation, it is a challenging NP-hard problem and optimal solutions can be found only for graphs with several thousands of vertices maximum [11]. In some cases, due to the enormous size of the real-life networks, even powerful meta-heuristic approaches are not enough, and external memory variations of them are required [1].

A clique can be though of as a set of objects having similar parameters. For example, a clique in the market graph represents a dense cluster of stocks whose prices exhibit a similar behavior over time. Therefore, dividing the market graph into a set of distinct cliques (clique partitioning) is a natural approach to classifying stocks (dividing the set of stocks into clusters of similar objects an approach to solve the clustering problem). It is worth mentioning that other clustering approaches can be applied well for studying financial markets [28].

Analyzing cliques can prove global properties of financial markets: for example, on the stock market of the US, large cliques despite very low edge density confirms the idea about the "globalization" of the market [14].

4.2 Independent Sets

A set of nodes S in a graph G is an independent set (stable set) if any two vertices in S are not adjacent. The Maximum Independent Set problem is to find the independent set of the maximum cardinality. It is obvious, that the Maximum Independent Set problem and the MCP are connected as to find the maximum independent set in the graph is the same as to find the maximum clique in the complementary graph.

For a market graph, a maximum independent set represents the largest "perfectly diversified" portfolio. Surprisingly, relatively small independent sets were found for the case of the US stock market [14]. But one can consider another interesting concept related to maximum independent sets that is finding a perfectly diversified portfolio containing any given stock. It turned out that, for every vertex in the market graph of the US, an independent set that contains this vertex was detected, and the sizes of these independent sets were almost the same, which means that it is possible to find a diversified portfolio containing any given stock using the market graph methodology [16].

The Graph Coloring Problem, that essentially represents the partitioning of the graph into a minimum number of independent sets, is another useful tool since it provides partitioning of a dataset represented by a graph into a number of clusters of "different" objects.

4.3 Connected Components

A connected component C is a subgraph of G such that it is connected, i.e. for any pair of vertices from C, there is a path from one vertex to another, and we can not add any vertex from the initial graph still having the connectivity property hold. Thus the initial graph can be decomposed into the set of non-overlapping connected components. And using either breadth-first search or depth-first search one can easily find all connected components of a graph in linear time.

As it has been already mentioned the real-life networks have a tendency to have a giant connectivity component [3]. For example, in the case of the call graph, it comprises around 80 % of all telephone numbers [1].

Studying behavior of a giant connected component over time one can get a useful information about the network. For instance, on a financial market, an increase in the giant component size from oldest to newest time period indicates the globalization tendency, just as in maximum clique size and edge density.

4.4 Minimum Spanning Trees

Given weighted connected graph G, a Minimum Spanning Tree (MST) is a subgraph of G such that: (a) the set of its vertices is the set of vertices of the initial graph; (b) it is connected graph; (c) it has no cycles; and (d) it has the minimally possible sum of edges. To find an MST one can use the well-known polynomial algorithms by Kruskal or by Prim. The Minimum Spanning Forest is the generalization of the concept for disconnected graphs where we find an MST for any connected component of a graph.

This problem has a number of applications in classical optimization and network theory: taxonomy, clustering analysis, Traveling Salesman Problem approximation, network design, etc. (see, for example, [47]).

In 1999 Mantegna [33] suggested using MSTs for obtaining a hierarchical structure of stocks on a financial market which allows one to estimate the importance of a particular stock. The approach was criticized for it uses only a small amount of information available and, therefore, the concept of the Planar Maximally Filtered Graph (PMFG) was introduced [43]. To find a PMFG is an NP-hard problem but it provides a useful subgraph comprising a lot of information, and which can be visualized easily since it is planar.

4.5   Robust Networks

In many case we are interested in construction/analysis of networks that will be robust against a potential failure of certain components.

One of the ways to tackle this problem is to first identify the most vulnerable components and estimate consequences of their removal. It can be formulated as the follows: given a graph $G(V, E)$ and an integer k, find a set of at most k elements, whose deletion minimizes the connectivity of the residual network. Here by elements we can mean nodes (arcs), paths, cliques, node subsets, etc. Also connectivity can have a number of interpretation: single/multiple commodity shortest path, or maximum flow, or minimum cost; number of pairwise connections, number of components, a largest component size, etc.

As an example, consider such a version of the Critical Node Detection Problem: given a graph $G(V, E)$ and an integer k, find a set of at most k nodes, whose deletion minimizes the pairwise connections of the residual network. It can be formulated as a Integer Programming model [7]:

$$
\begin{aligned}
\min \quad & \sum_{i,j \in V} u_{ij} \\
s.t. : \quad & u_{ij} + v_i + v_j \geq 1 \ \forall (i, j) \in E \\
& u_{ij} + u_{jk} - u_{ki} \leq 1 \ \forall (i, j, k) \in V^3 \\
& u_{ij} - u_{jk} + u_{ki} \leq 1 \ \forall (i, j, k) \in V^3 \\
& -u_{ij} + u_{jk} + u_{ki} \leq 1 \ \forall (i, j, k) \in V^3 \\
& \sum_{i \in V} v_i \leq k \\
& u_{ij} \in \{0; 1\} \ \forall (i, j) \in V^2 \\
& v_i \in \{0; 1\} \ \forall i \in V,
\end{aligned}
\tag{5}
$$

where $v_i = 1$ if node i is critical and 0, otherwise; $u_{ij} = 1$ if i and j are in the same component and 0, otherwise.

The problem is proven to be NP-hard in the general case for different elements: nodes (arcs), paths and cliques [7,23,45]. Notice that the selection of the connectivity measure is also crucial. Despite the fact that all these measures account for a disconnection level, using one over the other may lead to different critical elements.

This problem has a number of applications in many fiends: evacuation planning, fragmentation of terrorist organizations, epidemic contagion analysis and immunization planning, social network analysis (prestige and dominance), transportation (cross-dock and hub-and-spoke networks), marketing and customer services design, biomaterials and drugs design, etc. [41,42].

From the other hand, instead of analyzing vulnerable components we always can construct a robust network from the scratch. For example, consider the problem of finding a spanning k-tree of minimum weight in a complete weighted network [40]. A network constructed in such a way is robust in the sense that it remains connected even after destruction of $k-1$ its vertices. For other examples refer to [41,42].

5   Concluding Remarks

The proliferation of massive datasets brings with it a series of special computational and theoretical challenges [2].

The complexity of the analysis of these networks is determined by the following two facts: (1) most of the problems are NP-hard; (2) solving these problems will involve operating on massive amounts of data, which makes this process especially challenging.

To properly address constantly increasing volumes of considered problems, the development of new external algorithms and data structures, and new advances in parallel computing are needed. We need new conceptual approaches to global and local optimization such as quantum computing and local search techniques accounting structural and statistical properties of real-life networks.

Several theoretical aspects associated with models of real-life networks have yet to be studied. In particular, there is still no theoretical model that could describe all properties of these networks. Moreover, estimating the size of a maximum clique (independent set) or other more complicated structures in these models is an open question. Those are important problems, because, for example, the size of cliques and independent sets often provides useful information about the corresponding real-world network.

There is a need to create theoretical models characterizing local properties of real-world networks. Studying local properties of these networks implies the investigation of the structure of their small subgraphs.

Acknowledgments. This work is partially supported by the Laboratory of Algorithms and Technologies for Network Analysis, National Research University Higher School of Economics, Nizhny Novgorod, Russia.

References

1. Abello, J., Pardalos, P.M., Resende, M.: On maximum clique problems in very large graphs. In: Abello, J.M., Vitter, J.S. (eds.) External Memory Algorithms. DIMACS Series, vol. 50, pp. 119–130. AMS, Providence (1999)
2. Abello, J., Pardalos, P.M., Resende, M.G.S.: Handbook of Massive Data Sets. Kluwer Academic Publishers, Dordrecht (2002)

3. Aiello, W., Chung, F., Lu, L.: A random graph model for power law graphs. Exp. Math. **10**, 53–66 (2001)
4. Aiello, W., Chung, F., Lu, L.: Random evolution in massive graphs. In: Abello, J., Pardalos, P.M., Resende, M.G.C. (eds.) Handbook of Massive Data Sets, pp. 97–122. Kluwer Academic Publishers, Dordrecht (2002)
5. Albert, R., Jeong, H., Barabási, A.-L.: Internet: diameter of the world-wide web. Nature **401**, 130–131 (1999)
6. Albert, R., Barabási, A.-L.: Statistical mechanics of complex networks. Rev. Mod. Phys. **74**, 47–97 (2002)
7. Arulsevan, A., Commander, C.W., Elefteriadou, L., Pardalos, P.M.: Detecting critical nodes in sparse graphs. Comput. Oper. Res. **36**, 2193–2200 (2009)
8. Bansal, S., Khandelwal, S., Meyers, L.A.: Exploring biological network structure with clustered random networks. BMC Bioinform. **10**(405) (2009)
9. Barabási, A.-L., Albert, R.: Emergence of scaling in random networks. Science **286**, 509–512 (1999)
10. Barabási, A.-L., Albert, R., Jeong, H.: Scale-free characteristics of random networks: the topology of the world-wide web. Phys. A: Stat. Mech. Appl. **281**(14), 9–77 (2000)
11. Batsyn, M., Goldengorin, B., Maslov, E., Pardalos, P.M.: Improvements to MCS algorithm for the maximum clique problem. J. Comb. Optim. **27**, 397–416 (2014)
12. Bautin, G., Kalyagin, V., Koldanov, A., Koldanov, P., Pardalos, P.M.: Simple measure of similarity for the market graph construction. Comput. Manag. Sci. **10**, 105–124 (2013)
13. Boccaletti, S., Latora, V., Moreno, Y., Chavez, M., Hwang, D.-U.: Complex networks: structure and dynamics. Phys. Rep. **424**(4), 175–308 (2006)
14. Boginski, V., Butenko, S., Pardalos, P.M.: On structural properties of the market graph. In: Innovations in Financial and Economic Networks. Edward Elgar Publishers (2003)
15. Boginski, V., Butenko, S., Pardalos, P.M.: Modeling and optimization in massive graphs. In: Pardalos, P.M., Wolkowicz, H. (eds.) Novel Approaches to Hard Discrete Optimization, pp. 17–39. AMS, Providence (2003)
16. Boginski, V., Butenko, S., Pardalos, P.M.: Statistical analysis of financial networks. Comput. Stat. Data Anal. **48**, 431–443 (2005)
17. Bollobás, B.: Extremal Graph Theory. Academic Press, New York (1978)
18. Bollobás, B.: Random Graphs. Academic Press, New York (1985)
19. Bollobás, B., Riordan, O.M., Spencer, J., Tusnády, G.: The degree sequence of a scale-free random graph process. Random Struct. Algorithms **18**(3), 279–290 (2001)
20. Bollobás, B., Riordan, M.: Mathematical results on scale-free random graphs. In: Bornholdt, S., Schluster, H.G. (eds.) Handbook of Graphs and Networks: From the Genome to the Internet, pp. 1–34. Wiley-VCH, London (2003)
21. Bomze, I.M., Budinich, M., Pardalos, P.M., Pelillo, M.: The maximum clique problem. In: Du, D.-Z., Pardalos, P.M. (eds.) Handbook of Combinatorial Optimization, pp. 1–74. Kluwer Academic Publishers, Dordrecht (1999)
22. Broder, A., Kumar, R., Maghoul, F., Raghavan, P., Rajagopalan, S., Stata, R., Tomkins, A., Wiener, J.: Graph structure in the web. Comput. Netw. **33**, 309–320 (2000)
23. Dinh, T.N., Xuan, Y., Thai, M.T., Pardalos, P.M., Znati, T.: On new approaches of assessing network vulnerability: hardness and approximation. IEEE/ACM Trans. Netw. **20**(2), 609–619l (2012)

24. Google Official Blog. https://googleblog.blogspot.com/2008/07/we-knew-web-was-big.html
25. Hayes, B.: Graph theory in practice. Am. Sci. **88**, 9–13 (2000)
26. Iasemidis, L., Shiau, D., Sackellares, J., Pardalos, P.M.: Quadratic binary programming and dynamic system approach to determine the predictability of epileptic seizures. J. Comb. Optim. **5**, 9–26 (2001)
27. Iasemidis, L., Sackellares, J., Shiau, D., Chaovalitwongse, W., Carney, P., Principe, J., Yang, M., Yatsenko, V., Roper, S., Pardalos, P.M.: Seizure warning algorithm based on optimization and nonlinear dynamics. Math. Program. **101**(2), 365–385 (2004)
28. Kocheturov, A., Batsyn, M., Pardalos, P.M.: Dynamics of cluster structures in a financial market network. Phys. A: Stat. Mech. Appl. **413**, 523–533 (2014)
29. Korenkevych, D., Chien, J.-H., Zhang, J., Shiau, D.-S., Sackellares, C., Pardalos, P.M.: Small world networks in computational neuroscience. In: Pardalos, P.M., Du, D.-Z., Graham, R.L. (eds.) Handbook of Combinatorial Optimization, pp. 3057–3088. Springer, New York (2013)
30. Kumar, S.R., Raghavan, P., Rajagopalan, S., Tompkins, A.: Extracting large-scale knowledge bases from the web. In: Proceedings of the 25th International Conference on VLDB, pp. 639–650. Morgan Kaufmann Publishers (1999)
31. Kumar, S.R., Raghavan, P., Rajagopalan, S., Tompkins, A.: Trawling the web for emerging cyber communities. Comput. Netw. **31**(11–16), 1481–1493 (1999)
32. Kumar, S.R., Raghavan, P., Rajagopalan, S., Sivakumar, D., Tompkins, A., Upfal, E.: Stochastic models for the web graph. In: Proceedings of the 41st Annual Symposium on Foundations of Computer Science, pp. 57–65. IEEE Computer Society (2000)
33. Mantegna, R.N.: Hierarchical structure in financial markets. Eur. Phys. J. B **11**, 193–197 (1999)
34. Milgram, S.: The small-world problem. Psychol. Today **1**, 61–67 (1967)
35. Murre, J.M.J., Sturdy, D.P.F.: The connectivity of the brain: multi-level quantitative analysis. Biol. Cybern. **73**(6), 529–545 (1995)
36. Newman, M., Strogatz, S.H., Watts, D.J.: Random graphs with arbitrary degree distributions and their applications. Phys. Rev. E **64**, 026118 (2001)
37. Ostroumova, L., Ryabchenko, A., Samosvat, E.: Generalized preferential attachment: tunable power-law degree distribution and clustering coefficient. In: Bonato, A., Mitzenmacher, M., Prałat, P. (eds.) WAW 2013. LNCS, vol. 8305, pp. 185–202. Springer, Heidelberg (2013)
38. Serrano, M.A., Boguñá, M.: Tuning clustering in random networks with arbitrary degree distributions. Phys. Rev. E **72**(3), 036133 (2005)
39. Skidmore, F., Korenkevych, D., Liu, Y., He, G., Bullmore, E., Pardalos, P.M.: Connectivity brain networks based on wavelet correlation analysis in Parkinson fMRI data. Neurosci. Lett. **499**, 47–51 (2011)
40. Shangin, R.E., Pardalos, P.M.: Heuristics for minimum spanning k-tree problem. Procedia comput. sci. **31**, 1074–1083 (2014)
41. Thai, M., Pardalos, P.M.: Handbook of Optimization in Complex Networks: Theory and Applications. Springer Optimization and Its Applications. Springer, New York (2011)
42. Thai, M., Pardalos, P.M.: Handbook of Optimization in Complex Networks: Communication and Social Networks. Springer Optimization and Its Applications. Springer, New York (2011)
43. Tumminello, M., Aste, T., Matteo, T., Mantegna, R.N.: A tool for filtering information in complex systems. PNAS **102**, 10421–10426 (2005)

44. Volz, E.: Random networks with tunable degree distribution and clustering. Phys. Rev. E **70**(5), 056115 (2004)
45. Walteros, J.L., Pardalos, P.M.: A decomposition approach for solving critical clique detection problems. In: Klasing, R. (ed.) SEA 2012. LNCS, vol. 7276, pp. 393–404. Springer, Heidelberg (2012)
46. Watts, D.J., Strogatz, S.H.: Collective dynamics of 'small-world' networks. Nature **393**, 440–442 (1998)
47. Wu, B.Y., Chao, K.-M.: Spanning Trees and Optimization Problems. Taylor & Francis Group, London (2004)
48. Zhang, J., Xanthopoulos, P., Liu, C., Bearden, S., Uthman, B.M., Pardalos, P.M.: Real-time differentiation of nonconvulsive status epilepticus from other encephalopathies using quantitative EEG analysis: a pilot study. Epilepsia **51**(2), 243–250 (2010)

Models of Random Graphs
and Their Applications to the Web-Graph
Analysis

Andrei Raigorodskii[1,2,3,4]([✉])

[1] Lomonosov Moscow State University, Moscow, Russia
[2] Moscow Institute of Physics and Technology, Dolgoprudny, Russia
[3] Yandex, Moscow, Russia
[4] Institute of Mathematics and Computer Science,
Buryat State University, Ulan-Ude, Russia
mraigor@yandex.ru

Abstract. This course provides an overview of various models for random graphs and their applications to the Web graph. We start with the classical Erdős-Rényi model, then proceed with the most recent models describing the topology and growth of the Internet, social networks, economic network, and biological networks, and finally present several applications of these models to the problems of search and crawling.

Keywords: Web graphs · Random graphs

1 Introduction

Less than two decades ago most of ordinary people even did not know about Internet. Nowadays, Internet is a usual part of our everyday life; it provides us with a necessary means for information exchange and personal communication. Web search, e-mail, blogs, social networks, Internet messengers, etc. have become vital not only for younger generation. World Wide Web has demonstrated a tremendous growth during last years. One may even feel that it is completely chaotic since Web is really humongous and seems to be not controlled from a first glance. However, there are laws that govern Internet development. Study of these laws is attractive as well as study of laws of nature. This study had been started almost at the beginning of Internet advent. Now there is a whole discipline on the topic, which stays at the cross-roads between physics, mathematics and sociology.

The tutorial paper is organised as follows. In Sect. 2, we introduce basic objects of our studies and shortly discuss model-based approach. In Sect. 3, we summarise known properties of the real world web graphs. Then, in Sect. 6, we discuss Bollobás–Riordan model and its properties, and in Sects. 4 and 5 we introduce two preceding attempts, namely Erdős–Rényi model and Barabási–Albert "model". The clustering phenomenon and various definitions of clustering coefficients are

© Springer International Publishing Switzerland 2016
P. Braslavski et al. (Eds.): RuSSIR 2015, CCIS 573, pp. 101–118, 2016.
DOI: 10.1007/978-3-319-41718-9_5

discussed in Sect. 7. In Sect. 8 we examine Buckley-Osthus model of random graph in details since it overcomes several drawbacks of the previously introduced models. Section 9 presents conventional Page Rank definition and its possible alternatives. Since Buckley–Osthus model is not ideal for modelling clustering, we shortly discuss a new general class of models that bridge the gap in Sect. 10. Section 11 proposes so called "Fresh models", which aim to answer how to properly adjust preferential attachment mechanism by introduction of age factor for edges. In Sect. 12, we briefly discuss the directed preferential attachment model introduced by Bollobás, Borgs, Chayes and Riordan. In Sect. 13, we present so called "vertex copying" models. Finally, Sect. 14 concludes the tutorial.

2   Basic Objects

Let $G = (V, E)$ be a *real-world web-graph*, where V is a set of vertices (web-pages, set of web-sites, or set of web-hosts), and E is the set of all hyperlinks between the vertices (nodes). Sometimes multiple edges and even loops are allowed.

One may ask why do we need a model? There are many reasons, for example, it may help to properly adjust algorithms or find unexpected structures (news, spam, etc.) using classifiers learnt on some features coming from models.

And how to construct a model? There is a two-stage approach.

1. *Statistics.* First, find some statistical properties of web-graphs that would describe most accurately the real-world structures.
2. *Probability Theory.* Then, take a random element G which takes values in a set of graphs on n vertices and has such a distribution that w.h.p. (with high probability, i.e., with probability approaching 1 as $n \to \infty$) G has the same properties as the ones mentioned above.

3   Properties of the Real World Web Graphs

Properties of the real world web graphs were extensively studied in a number of papers. See, for example, Barabási–Albert [1–4], Watts–Strogatz [5], Newman [6], and many others in 90s–00s.

We shortly summarise the properties which are widely know so far below:

- Web-graphs are *sparse*, i.e., their numbers of edges (links) are proportional to their numbers of vertices;
- Web-graphs have a unique "giant" connected component;
- Every two vertices in the giant component are connected by a path of short length (5–6, 15–20 depending on what we mean by web-graph): $\mathrm{diam}\, G \approx 6$ (the rule of 6 handshakes);
- Web-graphs are robust when random vertices are destroyed (a giant component survives);
- Web-graphs are vulnerable to attacks onto hubs (many small components appear after a threshold is surpassed);

– The presence of many triangles in a web graph implies high clustering.
– The degree distribution is close to a power-law:

$$\frac{|\{v \in V : \deg v = d\}|}{n} \sim \frac{const}{d^{\gamma}},$$

where $\gamma \in (2, 3)$ depends on what we mean by web-graph.

4 Erdős–Rényi Model

Erdős–Rényi model was proposed and discussed in a series of papers [7–9].
    Let us introduce the model step by step:

– Fix $n \in \mathbb{N}$;
– Let $V_n = \{1, 2, \ldots, n\}$, $m = C_n^2$;
– Fix $p \in [0, 1]$;
– Consider all the edges $e_1, \ldots, e_m$ of the complete graph with vertices V_n;
– Take these edges independently, each with probability p;
– Denote the resulting random graph by $G(n, p)$.
– Now let $n \to \infty$. We may vary p depending on n.

Of course intuitively $G(n, p)$ does not reflect the reality. At least, it seems
unnatural to take the edges independently. However, if $p(n) = \frac{c}{n}$, then the
expected number of edges in $G(n, p)$ is $c\frac{n-1}{2}$, so that we get sparsity. And the
diameter is known to be small enough. But in this case the distribution of the
degrees is binomial with parameters $n - 1$, p, which is close, in turn, to Poisson
with the expectation c. It is not a power law at all!
    So, it would be great to explain somehow the power law phenomenon, not
loosing the sparsity and the diameter property.

5 Barabási–Albert "Model"

The original was proposed and studied in a series of papers [1–4].
    Let us again describe the model step by step below:

– Fix an initial graph G_0.
– Add new vertices (web sites) one by one.
– Any new vertex builds m edges with some previously existed vertices with
  probabilities proportional to the degrees of those vertices.
– This mechanism is called "preferential attachment": a new site prefers to
  establish a link to the existing sites according to their current popularity.

Obviously we have sparsity: n vertices and mn edges. Barabási and Albert
claim that this model results in, w.h.p., small diameter and — the most important — follows a power law.
    However, another pair of mathematicians, Bollobás and Riordan, criticised
not the idea but its realization.
    The criticism can be formalised as follows:

– First, of course the random graph depends greatly on the initial graph G_0;
– Second, Barabási and Albert do not say any word on how is distributed the set of m targets of a new vertex among the old ones;
– Finally, Bollobás and Riordan prove that, roughly speaking, one can organize the process (take a G_0 and a distribution of targets) in such a way that one gets, w.h.p., an arbitrary number of triangles in the resulting random graph.

In principle, it is good to have a class of models instead of just one model. The problem is that one can hardly understand from the papers by Barabási and Albert for which concrete model from this class the results are stated. Moreover, sometimes the results of different papers are contradictory, i.e. cannot be true simultaneously.

6 Bollobás–Riordan Model

6.1 Definition of the Model

The detailed description of Bollobás–Riordan model can be found in [9].

We should start with construction of a random graph G_m^n with n vertices and mn edges, $m \in \mathbb{N}$. Let $d_G(v)$ be the degree of a vertex v in a graph G. In what follows we consider two cases separately: $m = 1$ and $m > 1$.

Case $m = 1$.

Here, G_1^1 is a graph with one vertex v_1 and one loop.

Given G_1^{n-1} we can make G_1^n by adding vertex v_n and an edge from it to a vertex v_i, picked from $\{v_1, \ldots, v_n\}$ with probability

$$\mathbf{P}(i = s) = \begin{cases} \dfrac{d_{G_1^{n-1}}(v_s)}{2n-1} & 1 \le s \le n-1 \\ \dfrac{1}{2n-1} & s = n \end{cases}$$

Preferential attachment!

Case $m > 1$. Given G_1^{mn} we can make G_m^n by gluing $\{v_1, \ldots, v_m\}$ into v_1', $\{v_{m+1}, \ldots, v_{2m}\}$ into v_2', and so on.

The random graph G_m^n is certainly sparse. What is about other properties?

6.2 Diameter and Giant Components

The diameter of scale-free graphs is studied in [10].

Theorem 1 *(Bollobás, Riordan). If $m \ge 2$, then w.h.p. $\operatorname{diam} G_m^n \sim \frac{\ln n}{\ln \ln n}$.*

Great, since for real values of n, we get $\frac{\ln n}{\ln \ln n} \in [5, 15]$.
The robustness and vulnerability were studied in [11].

Theorem 2 *(Bollobás, Riordan). If $p \in (0, 1)$ and we make a random subgraph $G_{m,p}^n$ of the graph G_m^n by deleting its vertices independently each with probability p, then w.h.p. $G_{m,p}^n$ contains a connected component of size $\asymp n$.*

Great, since we have the robustness property.

Theorem 3 *(Bollobás, Riordan). If $c \in (0,1)$ and we make a random subgraph $G_{m,c}^n$ of the graph G_m^n by deleting its $[cn]$ first vertices, then for $c \leq (m-1)/(m+1)$, w.h.p. $G_{m,c}^n$ contains a connected component of size $\asymp n$, and for $c > (m-1)/(m+1)$, w.h.p. all the connected components of $G_{m,c}^n$ are of size $o(n)$.*

Great, since we have the vulnerability to attacks on the hubs.

6.3   Degree Distribution

Theorem 4 *(Bollobás et al. [12]). If $d \leq n^{1/15}$, then w.h.p.*

$$\frac{|\{v \in G_m^n : \deg v = d\}|}{n} \sim \frac{const(m)}{d^3}.$$

Great, since we get a power-law.

Not too great, since the exponent in the power-law is a bit different from the experimental ones ($\gamma \in (2,3)$).

Bad, since $d \leq n^{1/15}$, which is non-realistic.

The last problem is completely solved: analogue of Theorem 4 with an arbitrary d.

Tune the model somehow to get other exponents in the power-law?

7   Clustering

7.1   Various Definitions

Local Clustering Coefficient. Let $G = (V,E)$, $v \in V$ and N_v be the set of neighbours of v in G. Let $n_v = |N_v|$. If $n_v \geq 2$, then the local clustering coefficient of v is as follows [5]:

$$C_v = \frac{|\{\{x,y\} \in E : x,y \in N_v\}|}{C_{n_v}^2}.$$

Global Clustering Coefficient. The global clustering coefficient of G is defined as follows [6]:

$$T(G) = \frac{\sum\limits_{v \in V} C_{n_v}^2 C_v}{\sum\limits_{v \in V} C_{n_v}^2}.$$

Let $\sharp(H,G)$ be the number of copies of a graph H in a graph G. Then

$$T(G) = \frac{3\sharp(K_3,G)}{\sharp(P_2,G)},$$

where K_3 is a triangle and P_2 is a 2-path.

Average Local Clustering Coefficient. The average local clustering coefficient of G is

$$C(G) = \frac{1}{|V|} \sum_{v \in V} C_v.$$

In [6, 11], it was mentioned that either the average local or the global clustering coefficients is applied. However, it is not always clear which definition is used. In fact, the quantities $T(G)$ and $C(G)$ are quite different.

Let G be $K_{2,n-2}$ plus one edge between the vertices in the part of size 2. Then $C(G) \sim 1$, but $T(G) = \Theta\left(\frac{1}{n}\right)$.

Even though this difference is very important, there are many inaccuracies in the literature.

7.2   Experiments Versus Theory

Experimentally, $C(WWW)$ seems to be constant. However, there is no real data for $T(WWW)$. Newman asserts that $T(WWW)$ is constant as well without further explanations [6]. This is wrong, provided we do not take into account multiple edges and loops (which is widely done).

Theorem 5 *(Ostroumova, Samosvat [13]). If in a sequence $\{G_n\}$ of graphs, the degrees of the vertices follow a power law with exponent $\gamma \in (2, 3)$, then $T(G_n) \to 0$ as $n \to \infty$.*

Actually, Newman might be right, provided we do take into account multiple edges and loops.

Theorem 6 *(Ostroumova).[1] There exist sequences $\{G_n\}$ of multigraphs with loops, whose degrees of the vertices follow a power law with exponent $\gamma \in (2, 3)$ and, nevertheless, $T(G_n) \geq$ const as $n \to \infty$.*

The crucial question is, what is $T(G)$, if G has multiple edges and loops? There are many different definitions, but Newman does not mention this subtlety.

7.3   Clustering: The Bollobás-Riordan Model

Theorem 7 *(Bollobás, Riordan [14]). The expected value of $T(G_m^n)$ tends to 0 as $n \to \infty$: $\mathbf{E}(T(G_m^n)) \asymp \frac{\ln^2 n}{n}$.*

To calculate the value $\mathbf{E}(T(G_m^n))$, one needs to know the number of triangles and the number of 2-paths. Recall that $\sharp(H, G)$ is the number of copies of a graph H in a graph G.

A general and nice result was proved by Ryabchenko and Samosvat [15].

---

[1] The paper submitted to Internet Mathematics.

Theorem 8 *(Ryabchenko, Samosvat). For any H, $\mathbf{E}(\sharp(H, G_m^n)) \asymp n^{\sharp(d_i=0)} \cdot (\sqrt{n})^{\sharp(d_i=1)} \cdot (\ln n)^{\sharp(d_i=2)}$, where $\sharp(d_i = k)$ is the number of vertices of degree k in H.*

Theorem 9 *(Bollobás, Riordan). The expected value of $T(G_m^n)$ tends to 0 as $n \to \infty$: $\mathbf{E}(T(G_m^n)) \asymp \frac{\ln^2 n}{n}$.*

Theorem 10 *(Ryabchenko, Samosvat). For any H, $\mathbf{E}(\sharp(H, G_m^n)) \asymp n^{\sharp(d_i=0)} \cdot (\sqrt{n})^{\sharp(d_i=1)} \cdot (\ln n)^{\sharp(d_i=2)}$, where $\sharp(d_i = k)$ is the number of vertices of degree k in H.*

Theorem 10 agrees with Theorem 9: $\mathbf{E}(\sharp(K_3, G_m^n)) \asymp \ln^3 n$, $\mathbf{E}(\sharp(P_2, G_m^n)) \asymp n \ln n$.

By Theorem 10 the number of K_4 (and so on) is asymptotically constant, which is bad. Unfortunately, $C(G_m^n)$ also approaches 0, which is even worse. And, once again, $\gamma = 3$, which violates condition $\gamma \in (2, 3)$. So, let us tune the model and try to calculate again the number of **small subgraphs**.

8 Buckley–Osthus Model

8.1 Model Definition

One may ask which problems we had in the model of Bollobás–Riordan? Non-realistic exponent in the power-law, non-realistic clustering. It is possible to solve the first problem.

The following model is very close to the original one, but it has one important new parameter $a > 0$ called *initial attractiveness* of a vertex. An interested reader may also refer to [16,17].

Case $m = 1$. $H_{a,1}^1$ — graph with one vertex v_1 and one loop.

Given $H_{a,1}^{n-1}$ we can make $H_{a,1}^n$ by adding vertex v_n and an edge from it to a vertex v_i, picked from $\{v_1, \ldots, v_n\}$ with probability

$$\mathbf{P}(i = s) = \begin{cases} \dfrac{d_{H_{a,1}^{n-1}}(v_s) + a - 1}{(a+1)n - 1} & 1 \le s \le n - 1 \\[2ex] \dfrac{a}{(a+1)n - 1} & s = n \end{cases}$$

For $a = 1$, we get the model of Bollobás–Riordan.

Case $m > 1$. Given $H_{a,1}^{mn}$ we can make $H_{a,m}^n$ by gluing $\{v_1, \ldots, v_m\}$ into v_1', $\{v_{m+1}, \ldots, v_{2m}\}$ into v_2', and so on.

8.2 Buckley–Osthus Model: Degree Distribution

Theorem 11 *(Buckley, Osthus). If $d \le n^{1/(100(a+1))}$, then w.h.p.*

$$\frac{|\{v \in H_{a,m}^n : \deg v = d\}|}{n} \sim \frac{const(a, m)}{d^{a+2}}.$$

On the one hand this is a great result since now we can tune the model to get the expected exponent, but on the other hand it is bad since $d \le n^{1/(100(a+1))}$.

Anyway, the original problem is completely solved and there are many other great features of the model.

8.3 Buckley–Osthus Model: Second Degrees of Vertices

Let

$$d_2(t) = |\{\{i, j\} : i \neq t, j \neq t, \{i, t\} \in E(H^n_{a,1}), \{i, j\} \in E(H^n_{a,1})\}|.$$

So we calculate the number of edges of $H^n_{a,1}$ that are joined with a neighbour of a given vertex t.

Theorem 12 *(Ostroumova et al. [18]). W.h.p.*

$$\frac{|\{i = 1, \ldots, n : d_2(i) = d\}|}{n} \sim \frac{const(a)}{d^{a+1}}.$$

Fits quite well to the real data.

8.4 Buckley–Osthus Model: The Number of Edges Between Vertices of Given Degrees

Let $X_n(d_1, d_2)$ be the total number of edges between vertices of given degrees.
Subtleties with the definition!
Very important, since

$$|\{v \in V_n : \deg v = d\}| = \frac{1}{d} \sum_{d_1} X_n(d_1, d).$$

Theorem 13 *(Grechnikov [19]). W.h.p.*

$$\frac{X_n(d_1, d_2)}{n} \sim c(a, m) \left(\frac{(d_1 + d_2)^{1-a}}{d_1^2 d_2^2} \right).$$

8.5 Buckley–Osthus Model: "Power and Glory"

Theorem 14 *Theorem (Grechnikov). Let $d_1 \geq m$ and $d_2 \geq m$. Let $X = X_n(d_1, d_2)$. There exists a function $c_X(d_1, d_2)$ such that*

$$\mathbf{E}X_n(d_1, d_2) = c_X(d_1, d_2)n + O_{a,m}(1)$$

and

$$c_X(d_1, d_2) = \frac{\Gamma(d_1 - m + ma)\Gamma(d_2 - m + ma)}{\Gamma(d_1 - m + ma + 2)\Gamma(d_2 - m + ma + 2)} \times$$
$$\times \frac{\Gamma(d_1 + d_2 - 2m + 2ma + 3)}{\Gamma(d_1 + d_2 - 2m + 2ma + a + 2)} ma(a + 1) \frac{\Gamma(ma + a + 1)}{\Gamma(ma)} \times$$
$$\times \left(1 + \theta(d_1, d_2) \frac{(d_1 - m + ma + 1)(d_2 - m + ma + 1)}{(d_1 + d_2 - 2m + 2ma + 1)(d_1 + d_2 - 2m + 2ma + 2)} \right),$$

where

$$-4 + \frac{2}{1 + ma} \leq \theta(d_1, d_2) \leq a \frac{\Gamma(ma + 1)\Gamma(2ma + a + 3)}{\Gamma(2ma + 2)\Gamma(ma + a + 2)}.$$

Bollobás–Riordan Model: "Power and Glory"

Theorem 15 *Theorem (Grechnikov). If $d_1 < k, d_2 < k$ or $d_1 = d_2 = k$, then $X = 0$. If $d_1 \geq k, d_2 \geq k$ and $d_1 + d_2 \geq 2k + 1$, then the expected value of X is*

$$\mathbf{E}X = \frac{k(k+1)}{d_1(d_1+1)d_2(d_2+1)} \left(1 - \frac{C_{2k+2}^{k+1} C_{d_1+d_2-2k}^{d_1-k}}{C_{d_1+d_2+2}^{d_1+1}} \right) (2kt+1) -$$

$$- \sum_{n=1}^{k} \frac{C_{d_1+d_2-2n}^{d_1-n}}{d_1 d_2 C_{d_1+d_2}^{d_1}} \left(\frac{(2n)!}{n!(n+1)!} \frac{k+1}{2k} + [n=k] \frac{(2k)!}{2(k-1)!^2} \right) -$$

$$- [d_1 = k] \frac{(k-1)(k+1)}{2kd_2(d_2+1)} - [d_2 = k] \frac{(k-1)(k+1)}{2kd_1(d_1+1)} + O_{k,d_1,d_2} \left(\frac{1}{t} \right).$$

8.6 Buckley–Osthus Model: One More Evidence of Its Quality

Assume that the web-graph is governed by the Buckley–Osthus model. What is the most likely parameter a?

We may try to find an optimal a by comparing the reality with the fact that the number of vertices of degree d is close to d^{-2-a} (Grechnikov).

We may try to find an optimal a by comparing the reality with the fact that the number of edges between vertices of degree d_1 and d_2 is close to $(d_1 + d_2)^{1-a} d_1^{-2} d_2^{-2}$ (Grechnikov).

Assertion (Grechnikov et al. [20]). In both cases, the optimum is at the same $a \approx 0.27$.

8.7 Buckley–Osthus Model: An Application

We have seen that the model fits quite well the reality. How could we apply it?

Assume that a subgraph H of the real web-graph has been found by an algorithm. How could we check automatically whether this graph is "expected" or it probably represents an "unnatural" structure like a spam construction or an "explosion" (say, important news)?

An algorithm

- Calculate the total degrees of all the vertices of H (in the complete web-graph).
- For each pair of vertices of H calculate the expected number of edges between them using Step 1 and Grechnikov's results.
- Sum all the values found at Step 2.
- Compare the result of Step 3 with the real number of edges in H.

The difference between the real and the expected values can be used as a feature.

8.8 Buckley–Osthus Model: Clustering

Theorem 16 *Theorem (Eggemann, Noble [21]). If $a > 1$, then $\mathbf{E}(\sharp(K_3, H_{a,m}^n)) \asymp \ln n$ as $n \to \infty$.*

It's remarkable that for $a = 1$ (i.e., for the B–R model) we had $\ln^2 n$ instead of $\ln n$.

Theorem 17 *(Eggemann, Noble). If $a > 1$, then $\mathbf{E}(\sharp(P_2, H_{a,m}^n)) \asymp n$ as $n \to \infty$.*

Theorem 18 *(Eggemann, Noble). If $a > 1$, then $\mathbf{E}(T(H_{a,m}^n)) \asymp \frac{\ln n}{n}$ as $n \to \infty$.*

8.9 Buckley–Osthus Model: Small Subgraphs

Very recently Tilga has proved far-reaching generalizations and refinements of the theorems by Bollobás–Riordan, Ryabchenko–Samosvat, and Eggemann–Noble[2].

Theorem (Tilga). For any $a > 0$ and any fixed graph F, the order of magnitude of $\mathbf{E}(\sharp(F, H_{a,m}^n))$ is found.

The exact statement is quite cumbersome involving many parameters and cases. So we just give several most important and short enough corollaries.

Buckley–Osthus Model: Paths and Cliques.

Theorem (Tilga). Let $m \geq 2$ and $a < 1$, $\lambda = \frac{1}{a+1}$. Let P_l be a path of length l. Then for $n \to \infty$,

$$\mathbf{E}\left(\sharp(P_l, H_{a,m}^n)\right) = \begin{cases} n^{(2\lambda-1)k+1} \cdot \Theta(m^l) & \text{for } l = 2k, \\ n^{(2\lambda-1)k+1} \cdot \ln n \cdot \Theta(m^l) & \text{for } l = 2k+1. \end{cases}$$

Theorem (Tilga). Let K_k be a clique of size k, where $4 \leq k \leq m+1$. Then for $n \to \infty$,

$$\mathbf{E}\left(\sharp(K_k, H_{a,m}^n)\right) = \begin{cases} n^{1+(\lambda-1)(k-1)} \cdot \Theta(m^{C_k^2}) & \text{for } a < \frac{1}{k-2}, \\ \ln n \cdot \Theta(m^{C_k^2}) & \text{for } a = \frac{1}{k-2}, \\ \Theta(m^{C_k^2}) & \text{for } a > \frac{1}{k-2}. \end{cases}$$

For example, if $a = \frac{1}{3}$ (close to 0.27), then the number of K_5 is about $\log n$, and the number of K_4 is about $\sqrt[4]{n}$. Much more realistic than in the B–R model!

Buckley–Osthus Model: Cycles and Bicliques.

Theorem (Tilga). Let C_l be a cycle of length l. Then for $n \to \infty$,

$$\mathbf{E}\left(\sharp(C_l, H_{a,m}^n)\right) = \begin{cases} n^{(2\lambda-1)k} \cdot \Theta(m^l) & \text{for } l = 2k, \\ n^{(2\lambda-1)k} \cdot \ln n \cdot \Theta(m^l) & \text{for } l = 2k+1. \end{cases}$$

---

[2] Submitted to Izvestia Mathematics.

Theorem (Tilga). Let $K_{k,l}$ be a biclique with $2 \leq l \leq \min\{k, m\}$. Then for $n \to \infty$,

$$\mathbf{E}\left(\sharp(K_{k,l}, H_{a,m}^n)\right) = \begin{cases} n^{k(1+(\lambda-1)l)} \cdot \Theta(m^{kl}) & \text{for } a < \frac{1}{l-1}, \\ (\ln n)^k \cdot \Theta(m^{kl}) & \text{for } a = \frac{1}{l-1}, \\ \Theta(m^{kl}) & \text{for } a > \frac{1}{l-1}. \end{cases}$$

The number of bicliques shows how many communities are formed. For example, if $a = \frac{1}{3}$ (close to 0.27), then there are many $K_{k,4}$ and a lot of $K_{k,3}$, which was impossible in the B–R model (there are no vertices of degree < 3 in such graphs).

9 Various Page Rank Definitions

9.1 Classical "Google" PageRank

$G_n = (V_n, E_n)$ — web-graph.

Classical "Google" PageRank is the solution $\pi(n) = (\pi_i(n))_{i=1,\dots,|V_n|}$ to the system

$$\pi_i(n) = c \sum_{j \to i} \frac{\pi_j(n)}{\text{outdeg } j} + \frac{1-c}{|V_n|}, \quad i = 1, \dots, |V_n|.$$

The coordinates $\pi_i(n)$ follow a power-law similarly to the degrees (with another parameter of the law).

The same is true in a Barabási–Albert model.

Theorem 19 *(Avrachenkov [22]). For $i > 0$*

$$\mathbf{E}\pi_i(n) = \frac{1-c}{1+n} \left(\frac{1}{1+c} + \frac{c\Gamma\left(i + \frac{1}{2}\right)\Gamma\left(n + \frac{c}{2} + 1\right)}{(1+c)\Gamma\left(i + \frac{c}{2} + 1\right)\Gamma\left(n + \frac{1}{2}\right)} \right) \approx$$

$$\approx \frac{1-c}{1+n} \left(\frac{1}{1+c} + \frac{c}{1+c}\left(i + \frac{1}{2}\right)^{-\frac{1+c}{2}} \left(n + \frac{1}{2}\right)^{\frac{1+c}{2}} \right).$$

9.2 Weighted "Yandex" PageRank

$G_n = (V_n, E_n)$ — web-graph. In what follows, we omit n in the notation to make it less cumbersome. So $G = (V, E)$.

To each $i \in V$ we assign an l-dimensional vector $\mathbf{v}_i$ of features, and to each edge $i \to j$ in E we assign an m-dimensional vector $\mathbf{e}_{i,j}$ of features.

We have three parameters to be adjusted optimally.

Two main parameters.

- An l-dimensional vector ω, which gives us a weight of a vertex $i \in V$: the weight is $f(\omega, i) = (\omega, \mathbf{v}_i)$ — scalar product.

– An m-dimensional vector φ, which gives us a weight of an edge $i \to j$ in E: the weight is $g(\varphi, i \to j) = (\varphi, \mathbf{e}_{i,j})$.

The third parameter $c \in (0, 1)$ will appear on the next slide.

Main definition.

Weighted PageRank is a vector π — the solution to a system

$$\pi_i = c \frac{\sum\limits_{j \to i} \pi_j g(\varphi, j \to i)}{\sum\limits_{h: \, j \to h} g(\varphi, j \to h)} + (1 - c) \frac{f(\omega, j)}{\sum\limits_{k \in V} f(\omega, k)}.$$

Main problem.

Let S be a measure of the difference between our PageRank $\pi(\omega, \varphi, c)$ and some estimates assigned to each document (to each $i \in V$) according to a given search query (e.g., S is the standard deviation). Find

$$(\omega_0, \varphi_0, c_0) = \mathrm{argmin}_{\omega, \varphi} S(\pi(\omega, \varphi, c), \text{vector of estimates}).$$

Note that this work is in progress and recent advances has been made in cooperation with Nesterov and his research group [23].

10   A New General Class of Models

10.1   Basic Deifnitions

Buckley–Osthus is not ideal for clustering. Can one do anything to improve it?

Many multiparametric models.

A break-through is due to Ryabchenko et al. [24].

The PA-class. Let G_m^n ($n \geq n_0$) be a graph with n vertices $\{1, \ldots, n\}$ and mn edges obtained as a result of the following random graph process. We start at the time n_0 from an arbitrary graph $G_m^{n_0}$ with n_0 vertices and mn_0 edges. On the $(n + 1)$-th step ($n \geq n_0$), we make the graph G_m^{n+1} from G_m^n by adding a new vertex $n + 1$ and m edges connecting this vertex to some m vertices from the set $\{1, \ldots, n, n + 1\}$. Denote by d_v^n the degree of a vertex v in G_m^n. Assume that for some constants A and B the following conditions are satisfied:

The PA-class conditions.

$$\mathbf{P}\left(d_v^{n+1} = d_v^n \mid G_m^n\right) = 1 - A\frac{d_v^n}{n} - B\frac{1}{n} + O\left(\frac{(d_v^n)^2}{n^2}\right), \; 1 \leq v \leq n, \quad (1)$$

$$\mathbf{P}\left(d_v^{n+1} = d_v^n + 1 \mid G_m^n\right) = A\frac{d_v^n}{n} + B\frac{1}{n} + O\left(\frac{(d_v^n)^2}{n^2}\right), \; 1 \leq v \leq n, \quad (2)$$

$$\mathbf{P}\left(d_v^{n+1} = d_v^n + j \mid G_m^n\right) = O\left(\frac{(d_v^n)^2}{n^2}\right), \; 2 \leq j \leq m, \; 1 \leq v \leq n, \quad (3)$$

$$\mathbf{P}(d_{n+1}^{n+1} = m + j) = O\left(\frac{1}{n}\right), \; 1 \leq j \leq m. \quad (4)$$

For $A = 1/2$, $B = 0$, we get Bollobás–Riordan.

For $A = 1/(1 + a)$, $B = ma/(1 + a)$, we get Buckley–Osthus.

10.2   A New General Class of Models: Results and Solutions

Theorem (Ostroumova, Ryabchenko, Samosvat) W.h.p.

$$\frac{|\{v \in G_m^n : \deg v = d\}|}{n} \sim \frac{const(A, B, m)}{d^{1+1/A}}.$$

Theorem (Ostroumova, Ryabchenko, Samosvat).

- If $2A < 1$ then w.h.p. $T(n) \sim c(A, B, m)$,
- If $2A = 1$ then w.h.p. $T(n) \sim \frac{c'(A,B,m)}{\ln n}$,
- If $2A > 1$ then for any $\varepsilon > 0$ w.h.p. $n^{1-2A-\varepsilon} \le T(n) \le n^{1-2A+\varepsilon}$.

Great, since in the first case, we have a constant clustering together with power-law!

Open questions and questions under solution.

- (Ostroumova, Krot) [25][3] The local clustering coefficient is constant in n in the new models and is proportional to $1/d$ in the degree d.
- What's with "Google", "Yandex" and other PageRanks in the new models?

11   "Fresh" Models

Another problem of the preferential attachment models: the older is a page, the greater is the probability that it will gain more and more citations. That's not completely realistic, for we have a large part of the Internet containing news and other media.

Need to use an age factor for each page.

Samosvat and Ostroumova [26] proposed a class of such models.

Let's construct a sequence of random graphs $\{G_n\}$. For that, consider an integer constant m (vertex outdegree), an integer function $N(n)$, and a sequence of mutually independent random variables $\zeta_1, \zeta_2, \ldots$ with some given distribution taking positive values.

First, take 2 vertices and 1 edge between them (graph $\tilde{G}_2^n$). The first 2 vertices have "inherent qualities" $q(1) = \zeta_1, q(2) = \zeta_2$. At the $t+1$-th step ($2 \le t \le n-1$) one vertex and m edges are added to $\tilde{G}_t^n$:

$$q(t + 1) = \zeta_{t+1}, \quad \mathbf{P}(t + 1 \to i) = \frac{\text{attr}_t(i)}{\sum_{j=1}^{t} \text{attr}_t(j)},$$

where

$$\text{attr}_t(i) = (1 \text{ or } q(i)) \cdot (1 \text{ or } d_t(i)) \cdot \left(1 \text{ or } \mathbf{I}[i > t - N(n)] \text{ or } e^{-\frac{t-i}{N(n)}}\right)$$

and $d_t(i)$ is the degree of the vertex i in $\tilde{G}_t^n$.

Let $G_n = \tilde{G}_n^n$.

---

[3] Accepted to WAW.

Theorem 20 *(Samosvat, Ostroumova). The most realistic case is when $attr_t(i) = q(i) \cdot e^{-\frac{t-i}{N(n)}}$ and ζ_i have a power law (say, Pareto) distribution. In this case, not only w.h.p. the fraction of the number of vertices with a given degree follows a power law with a tunable exponent, but also the quantity $e(T)$, which is the fraction of edges that connect vertices with age difference greater than T, decreases exponentially, right as in the reality.*

In fact, the model is used in Yandex to substantially improve the crawling algorithm.

12   Bollobás–Borgs–Chayes–Riordan Model

12.1   Model Definition

The model was proposed in [27].

Let us Construct a random *directed* graph $G(t)$.

The parameters are $\delta_{\text{in}} \geq 0$, initial attractiveness, $\delta_{\text{out}} \geq 0$, initial will to produce hyperlinks, $\alpha \geq 0$, $\beta \geq 0$, and $\gamma \geq 0$ s.t. $\alpha + \beta + \gamma = 1$.

The model $G(1)$ — graph with one vertex v_1 and one loop.

Given $G(t-1)$ and assuming it has $n(t-1)$ vertices (now, this value is random!) we can make $G(t)$ in three different ways.

1. With probability α a new vertex is added and it sends an edge to some vertex v among the previous $n(t-1)$ vertices with probability

$$\frac{\text{indeg}\, v + \delta_{\text{in}}}{t - 1 + \delta_{\text{in}} n(t-1)}. \tag{5}$$

2. With probability γ a new vertex is added and it receives an edge from some vertex v among the previous $n(t-1)$ vertices with probability

$$\frac{\text{outdeg}\, v + \delta_{\text{out}}}{t - 1 + \delta_{\text{out}} n(t-1)}. \tag{6}$$

3. With probability β no new vertices appear, and we pick two vertices v and w among the existing ones independently: v with probability as in (5) and w with probability as in (6). We draw an edge (v, w) (a loop may appear).

12.2   Bollobás–Borgs–Chayes–Riordan Model: The Degrees

Theorem 21 *(Bollobás–Borgs–Chayes–Riordan). Let a natural $i \geq 1$ be fixed. Let $x_i(t)$ $(y_i(t))$ be the number of vertices in $G(t)$ that have indegree (outdegree) i. Let*

$$c_1 = \frac{\alpha + \beta}{1 + \delta_{in}(\alpha + \gamma)}, \quad c_2 = \frac{\beta + \gamma}{1 + \delta_{out}(\alpha + \gamma)}.$$

Then there exist p_i, q_i, which are constant for any i, and there exist φ_i, ψ_i, which are functions depending on t and which are, for any given i, equal to $o(t)$, such that w.h.p. for each t, $x_i(t) = p_i t + \varphi_i(t)$, $y_i(t) = q_i t + \psi_i(t)$. Moreover, if $\alpha \delta_{in} + \gamma > 0$ and $\gamma < 1$, then there exists a constant $C_{in} > 0$ such that as $i \to \infty$

$$p_i \sim C_{in} i^{-1-\frac{1}{c_1}}.$$

If, in turn, $\gamma \delta_{out} + \alpha > 0$ and $\alpha < 1$, then there exists a constant $C_{out} > 0$ such that as $i \to \infty$

$$q_i \sim C_{out} i^{-1-\frac{1}{c_2}}.$$

13   Copying Models

13.1   Model Definitions

Parameters: natural $d \geq 1$ and $\alpha \in (0, 1)$.

The idea: sometimes new vertices choose random targets, but sometimes they copy hyperlinks from the sites that are of interest for them.

The aim was to explain both power law and many communities (bipartite subgraphs).

Kumar et al. [28]

Take a G_0 with all the degrees at least d. Assume that a graph G_t is already produced. Add one vertex to it and d edges in the following way. Take a random p from the set of existing vertices (uniform distribution). It has at least d neighbours: $p_1, \ldots, p_d, \ldots$ Now start producing new edges (independently). The ith edge goes with probability α to a random vertex among the existing ones (uniform distribution), and it goes with probability $1 - \alpha$ to p_i.

13.2   Copying Models: The Degrees

Let $N_{t,r}$ be the number of vertices of degree r in the graph with t vertices.

Theorem 22 (Kumar, Raghavan, Rajagopalan, Sivakumar, Tomkins, Upfal).
Let $d = 1$, $r > 0$. Let $P_r = \lim\limits_{t \to \infty} \frac{\mathbf{E} N_{t,r}}{t}$. This value is correctly defined and is given by

$$P_r = P_0 \prod_{i=1}^{r} \frac{1 + \alpha/(i(1 - \alpha))}{1 + 2/(i(1 - \alpha))}.$$

Roughly,

$$P_r = \Theta\left(r^{-\frac{2-\alpha}{1-\alpha}}\right).$$

One may find α to get an arbitrary parameter of the power law greater than 2!

13.3   Copying Models: Communities

Let $K_{i,j}$ be the complete bipartite graph with part sizes i, j.

Theorem 23 *(Kumar, Raghavan, Rajagopalan, Sivakumar, Tomkins, Upfal). Let d be a fixed natural number and $\alpha \in (0,1), t \to \infty, i \leq \ln t$. Then there exists a constant $c = c(d)$ such that*

$$\mathbf{P}\left(\sharp(K_{i,d}, G_t) \geq cte^{-i}\right) \to 1, \quad t \to \infty.$$

For example, if $d = 10, i = \ln t/2$, then w.h.p. the number of communities of size i citing some 10 sites is at least $\sqrt{t}$.

Even much more impressive than in the Buckley–Osthus model!

14   Conclusion

We have studied many different aspects of random graph models, which related to Internet studies and its theoretical properties.

And this is not yet a complete survey and there are many omitted questions since mainly our recent results are considered in this paper.

However, we are going to continue our study of other aspects of random graphs models. Thus, there are many other characteristics that are studied in the domain like assortativity, disassortativity, centrality, betweenness, etc.

There are not only the models of preferential attachment of graph formation, but game-theoretic ones (see book [29], for example). Those models demonstrate slightly worse behaviour, but they definitely have good prospects from mathematical viewpoint and are used in a range of applications, for example, in social sciences.

An interesting alternative to the considered models is so called geometric random graphs [30] where, in the basic setting, such random graphs are built by dropping n points randomly uniformly into the unit square and the edges can connect any two points at distance no longer than r. In such application as wireless communications, epidemiology, astronomy, the Internet and some others, geometric random graphs is a more realistic model than the model of Erdos and Renyi.

An interested reader may further refer to the books of Newman [31], Drogovtsev [32], and Durrett [33] since they cover various aspects of random graphs and complex networks.

Acknowledgments. The author was supported by Russian Foundation for Basic Research, grant no. 15-01-00350, and by grant NSH-2964.2014.1 for support of leading scientific schools.

References

1. Barabási, A.L., Albert, R.: Emergence of scaling in random networks. Science **286**, 509 (1999)
2. Barabási, A., Albert, R., Jeong, H.: Scale-free characteristics of random networks: the topology of the world-wide web. Phys. A: Stat. Mech. Appl. **281**(1–4), 69–77 (2000)
3. Albert, R., Jeong, H., Barabási, A.L.: The diameter of the world wide web. Nature **401**, 130–131 (1999)
4. Albert, R., Jeong, H., Barabási, A.L.: Error and attack tolerance of complex networks. Nature **406**(6794), 378–382 (2000)
5. Watts, D.J., Strogatz, S.H.: Collective dynamics of 'small-world' networks. Nature **393**(6684), 409–410 (1998)
6. Newman, M.E.J.: The structure and function of complex networks. SIAM Rev. **45**(2), 167–256 (2003)
7. Erdős, P., Rényi, A.: On random graphs, I. Publ. Math. (Debr.) **6**, 290–297 (1959)
8. Erdős, P., Rényi, A.: On the evolution of random graphs. In: Publication of the Mathematical Institute of the Hungarian Academy of Sciences, pp. 17–61 (1960)
9. Bollobás, B.: Random Graphs, 2nd edn. Cambridge University Press, Cambridge (2001)
10. Bollobás, B., Riordan, O.: The diameter of a scale-free random graph. Combinatorica **24**(1), 5–34 (2004)
11. Bollobás, B., Riordan, O.: Robustness and vulnerability of scale-free random graphs. Internet Math. **1**(1), 1–35 (2003)
12. Bollobás, B., Riordan, O., Spencer, J., Tusnády, G.E.: The degree sequence of a scale-free random graph process. Random Struct. Algorithms **18**(3), 279–290 (2001)
13. Ostroumova Prokhorenkova, L., Samosvat, E.: Global clustering coefficient in scale-free networks. In: Bonato, A., Graham, F.C., Prałat, P. (eds.) WAW 2014. LNCS, vol. 8882, pp. 47–58. Springer, Heidelberg (2014)
14. Bollobás, B.: Mathematical results on scale-free random graphs. In: Handbook of Graphs and Networks, Wiley, pp. 1–37 (2003)
15. Ryabchenko, A., Samosvat, E.: On the number of subgraphs of the Barabási-Albert random graph. Izv. Math. **76**(3), 607–625 (2012)
16. Dorogovtsev, S.N., Mendes, J.F.F., Samukhin, A.N.: Structure of growing networks with preferential linking. Phys. Rev. Lett. **85**, 4633–4636 (2000)
17. Buckley, P.G., Osthus, D.: Popularity based random graph models leading to a scale-free degree sequence. Discret. Math. **282**(13), 53–68 (2004)
18. Kupavskii, A., Ostroumova, L., Shabanov, D., Tetali, P.: The distribution of second degrees in the Buckley–Osthus random graph model. Internet Math. **9**(4), 297–335 (2013)
19. Grechnikov, E.: An estimate for the number of edges between vertices of given degrees in random graphs in the bollobás-riordan model. Moscow J. Comb. Number Theory **1**(2), 40–73 (2011)
20. Zhukovskiy, M., Vinogradov, D., Pritykin, Y., Ostroumova, L., Grechnikov, E., Gusev, G., Serdyukov, P., Raigorodskii, A.: Empirical validation of the Buckley-Osthus model for the web host graph: degree and edge distributions. In: Proceedings of the 21st ACM International Conference on Information and Knowledge Management, CIKM 2012, New York, NY, USA, pp. 1577–1581. ACM (2012)

21. Eggemann, N., Noble, S.: The clustering coefficient of a scale-free random graph. Discret. Appl. Math. **159**(10), 953–965 (2011)
22. Avrachenkov, K., Lebedev, D.: PageRank of scale-free growing networks. Internet Math. **3**(2), 207–231 (2007)
23. Bogolubsky, L., Dvurechensky, P., Gasnikov, A., Gusev, G., Nesterov, Y., Raigorodskii, A., Tikhonov, A., Zhukovskii, M.: Learning supervised PageRank with gradient-free optimization methods, p. 11, November 2014
24. Ostroumova, L., Ryabchenko, A., Samosvat, E.: Generalized preferential attachment: tunable power-law degree distribution and clustering coefficient. In: Bonato, A., Mitzenmacher, M., Prałat, P. (eds.) WAW 2013. LNCS, vol. 8305, pp. 185–202. Springer, Heidelberg (2013)
25. Krot, A., Prokhorenkova, L.O.: Local clustering coefficient in generalized preferential attachment models, July 2015
26. Ostroumova, L., Samosvat, E.: Recency-based preferential attachment models, June 2014
27. Bollobás, B., Borgs, C., Chayes, J., Riordan, O.: Directed scale-free graphs. In: Proceedings of the Fourteenth Annual ACM-SIAM Symposium on Discrete Algorithms, SODA 2003, Philadelphia, PA, USA, pp. 132–139. Society for Industrial and Applied Mathematics (2003)
28. Kumar, R., Raghavan, P., Rajagopalan, S., Sivakumar, D., Tomkins, A., Upfal, E.: Stochastic models for the web graph. In: Proceedings of the 41st Annual Symposium on Foundations of Computer Science, pp. 57–65 (2000)
29. Goyal, S.: Connections: An Introduction to the Economics of Networks. Princeton University Press, Princeton (2007)
30. Penrose, M.: Random Geometric Graphs. Oxford University Press, Oxford (2003)
31. Newman, M., Barabási, A.L., Watts, D.J.: The Structure and Dynamics of Networks. Princeton Studies in Complexity. Princeton University Press, Princeton (2006)
32. Dorogovtsev, S.: Lectures on Complex Networks. Oxford University Press Inc., New York (2010)
33. Durrett, R.: Random Graph Dynamics. Cambridge Series in Statistical and Probabilistic Mathematics. Cambridge University Press, New York (2006)

Young Scientist Conference Papers

Who Are My Ancestors? Retrieving Family Relationships from Historical Texts

Julia Efremova[1(✉)], Alejandro Montes García[1],
Alfredo Bolt Iriondo[1], and Toon Calders[1,2]

[1] Eindhoven University of Technology, Eindhoven, The Netherlands
{i.efremova,a.montes.garcia,a.bolt.iriondo}@tue.nl
[2] Université Libre de Bruxelles, Brussels, Belgium
toon.calders@ulb.ac.be

Abstract. This paper presents an approach for automatically retrieving family relationships from a real-world collection of Dutch historical notary acts. We aim to retrieve relationships like *husband - wife*, *parent - child*, *widow of*, etc. Our approach includes person names extraction, reference disambiguation, candidate generation and family relationship prediction. Since we have a limited amount of training data, we evaluate different feature configurations based on the n-gram analysis. The best results were obtained by using a combination of bi-grams and tri-grams of words together with the distance in words between two names. We evaluate our results for each type of the relationships in terms of precision, recall and $f - score$.

1 Introduction

Extraction of characteristics from the text is one of the main tasks in text mining. Structured information retrieved from the text can be used for different purposes, for instance, documents classification, filtering emails, finding key words, etc. Having extracted person names and their relationships, makes available a large amount of personal information. Previous research showed good results in automated extraction of skills from job applicants to make job application process more efficient [10].

Family relationships is a special type of person relationships. It is an important step in linking persons across different genealogical documents and sources [6,7]. As an example, consider a couple *Martinus de Jager* and *Hendrina Jacobs* who married in *1888*. The information about their marriage is recorded in a civil register. Two years later *Martinus* and *Hendrina* are mentioned as husband and wife in a notary act because they bought a house. Having extracted the *husband-wife* relationship between them can help to link this notary act to the marriage certificate of the mentioned couple or the birth certificates of their children where *Martinus* and *Hendrina* are mentioned as parents.

As such, extraction of family relationships from text documents is important for population reconstruction which is a key element in the genealogical research

© Springer International Publishing Switzerland 2016
P. Braslavski et al. (Eds.): RuSSIR 2015, CCIS 573, pp. 121–129, 2016.
DOI: 10.1007/978-3-319-41718-9_6

and social studies. Family relationships extraction can be also used in text classification. For instance, regarding a collection of notary acts, it is much easier to distinguish between an inheritance act and a purchase activity because the first type contains many family relationships and the second usually not.

In this paper we focus on the extraction of family relationships (FR) from historical notary acts which is a challenging text mining problem. We deal with identification of FRs objects, FRs key words abbreviations, name variation in the text or implicit relationships. We identify main components such as extraction of named entities (person names in our case), name disambiguation if a person is mentioned more than once in the document, and relationship prediction.

Our contributions can be summarized as follows:

– We propose a framework that allows the retrieval of family relationships extraction from the text with minimal available training data.
– We report results of n-gram analysis that are used as a feature configuration technique.
– We provide a training collection that consists of labeled notary acts and relationships pairs to the research community.

The remainder of this paper has the following structure. In Sect. 2 we discuss related work. In Sect. 3 we describe the data collection. In Sect. 4 we present a general process of family relationship extraction. In Sect. 5 we set up the experiments and present the results. Finally, we make a conclusion in Sect. 6.

2   Related Work

In this section we discuss the related work on family relationships extraction. Santos et al. [15] apply a rule-based approach in order to extract family relationships. This method contains of 99 different rules which allow to identify and to classify family relationships. Their obtained f-score varies from 29 % to 36 % and the designed rule based patterns are restricted to the Spanish text. Makazhanov [12] extracts FR networks from literary novels. He uses literature narratives and considers utterances in the text which are attributed into different categories: quotes, apparent conversations, character tri-gram and others. Then the FR prediction is done by using a Naive Bayes classifier and it is evaluated on the book of Jane Austen *Pride and Prejudice*. Kokkinakis and Malm [11] describe an unsupervised approach to extract interpersonal relations between identified person entities from Swedish prose. Recently Collovini et al. [3] designed a process for the extraction of any types of relations between named entities for Portuguese text in the domain of organisations. They apply statistical modelling with different feature combinations. Bird et al. [2] describe relationship extraction based on regular expressions and pattern features. However, their method requires a dictionary of named entities. For instance, they use *in* pattern to find the location of organisations: *[ORG: Bastille Opera] 'in' [LOC: Paris]*. Mintz et al. [13] propose an approach for relation extraction from the text that does not require labelled data. They focus on identifying pairs, for example, the *person-nationality* relation which holds between person entities and nationality entities. In our work we aim to identify triples ($person_1$, *family relationship*, $person_2$). Based on the previous work applied to different languages and application domains we design our own framework for FR extraction from historical documents.

We also mention a number of general work available in text classification. One of the main surveys in text classification (TC) was published by Sebastiani [16] where the author described main TC steps such as document indexing, dimensionality reduction, key term selection, learning a classifier and evaluation. Ikomomakis [9] extended later his work and made a summary of the available TC technique which contain a number of relevant cited references. One of the recent survey belong to Aggarwal and Zhai [1]. They described main TC components together with state-of-the-art solutions from data mining, machine learning and information retrieval.

3   Data Description

The dataset used is a subset of the one described in our previous work [5] and it is available on the web[1]. More specifically we give the description of the annotated input collection in Sect. 5.1. The collection contains historical notary acts such as: property transfer, sale, inheritance, public sale, obligation, declaration, partition of inheritance, resolution, inventory and evaluation for a period of around 500 years. Many of the notary acts contain information about people and family relationships between them. Since the documents belong to a time period that was many years ago sometimes they are the only sources of information regarding the population and family relations of that period. Thus, we need an efficient technique to extract person entities and their relationships.

Below is an example of a notary act that has the *husband-wife* relationship (the person names are underlined and relationships are in bold):

> Dit document certificeert: <u>Martinus de Jager</u> en **zijn vrouw** <u>Hendrina Jacobs</u>, verklaren afstand te doen van alle rechten van de akte van koop en verkoop van 02/10/1906, opgemaakt voor notaris van Breda, ten behoeve van <u>Martinus van Doorn</u>, winkelier te Uden.
>
> *This document certifies:* <u>*Martinus de Jager*</u> *and* **his wife** <u>*Hendrina Jacobs*</u>, *declare to waive all rights of the act of sale and purchase of 02/10/1906, registered at the notary Breda, as beneficiary* <u>*Martinus van Doorn*</u>, *shopkeeper in Uden.*

The average length of the documents is 70 words, although there exist some documents with up to 1,000 words. The overall collection contains around 115,000 notary acts with dates ranging from *1433* to *1920*. However, the majority of documents belongs to the period *1650-1850*. Different time period of documents and different documenting standards make the task of family relationship extraction very difficult.

---

[1] http://goo.gl/leibR9.

4   Family Relationship Extraction

In this section we discuss the following steps of the FR retrieval process: *raw data pre-processing, name extraction, name disambiguation, candidate pair generation, feature generation* and *classification*.

4.1   Name Extraction

To extract person names from notary acts we use a collection of Dutch first and last names obtained from the website of Meertens Institute[2] available in Dutch only. It contains around $115,000$ different last names, $18,000$ male and $26,000$ female first names. We use this database as a name dictionary. Although the name dictionary is large, we can not apply it directly and tag all first and last names in the text. Some name variations might be missed. To avoid this situations we designed our own name extraction that proceeds in three steps.

In the first step, we define a set of labels {*FN, LN, I, P, CAP, O*} in which *'FN'* and *'LN'* stand for first and last names respectively, the tag *'I'* refers to a name initial (one letter followed by a dot like *'W.'* instead of *'Willem'*), *'P'* is a name prefix like *van, der, de*, *'CAP'* corresponds to other words that start from a capital letter and *'O'* indicates that there is no name descriptor. We assign an appropriate label to every word in the document in two iterations. We first tag first names and last names using the name dictionary, then we tag initials, name prefixes, words that start from a capital letter and other words that are not tagged yet.

In the second step we design name patterns using regular expressions. The phrase in the text is extracted as a name if it meets the requirements of a name pattern. Table 1 shows the three main name patterns that we used to specify a name phrase. The first name pattern corresponds to the situation when at least one first name exists in the dictionary. A last name is optional in this case and can be tagged as *'LN'* or *'CAP'*. If the last name does not exist in the dictionary we consider a word after the first name that starts from a capital letter as the last name. Between first and last name, initials or a name prefix may appear. This rule allows us to extract a single first name and full names at the same time. The second expression in Table 1 finds names that start from initials followed by the last name which can be tagged again as *'LN'* or *'CAP'*. The third expression requires a last name tag whereas the first name can be labelled with *'FN'* or *'CAP'* tags. We illustrate the process of labelling words and finding name patterns in Fig. 1.

In the third step we make name disambiguation and merge the same names into one. Name disambiguation is a necessary step in case when a person is mentioned multiple times. However, it is not a fully person resolution methods. The goal of this paper is to identify pair of persons in a documents and predict

---

[2] http://www.meertens.knaw.nl/nvb/.

Table 1. The grammar that specifies a name pattern

No	Name pattern
1	{<CAP>?<FN>+<I>?<P>?<LN|CAP>?}
2	{(<I>)+<FN>?<I>?<LN|CAP>+}
3	{<FN|CAP>+<P>?+<LN>}

their relationships. In order to do it we make a pairwise relationship prediction between two names. The extraction of person entities is not in the scope of this paper and is considered as a part of future work. The problem is that different persons can have the same name, for instance *Hendrina Jacobs and her daughter Hendrina Jacobs* and the same person can be mentioned differently: *Hendrina Jacob* and *H. Jacobs*. Our hypothesis is that family extraction technique can be a component of person resolution process. That is why we deal with names but not with different persons.

This simple technique extracts names with high accuracy, efficiently deals with abbreviations in them: *W.G. van Oijen* or *Jan J. Beckers* and distinguishes person names from other information and location in the text. For instance, compare the name *Jan van Erp* and the phrase *Kerk van Erp*[3]. The proposed name extraction technique is able to distinguish these two situations from each other.

4.2   General Approach

To retrieve FRs we start with data preprocessing and remove from the raw data all non-alphabetical symbols except punctuation marks. In the next step we extract person names from notary acts and perform name disambiguation as described in Sect. 4.1. For every pair of names extracted from the same notary act, we construct a candidate pair and create a feature vector for that pair. The feature vectors are constructed as follows. We consider all words between the two names in a pair and also two words before the first name and two words after the last name as illustrated in Fig. 1. Thus, for each candidate pair we identify a set of words called *tokens*.

We compute *term frequency* for each token in a candidate pair. The output of the feature extraction step is a set of numerical features. The created vocabulary is large. We experiment with different sets of features: *bi-grams* of words, *tri-grams* of words, a combination of *bi-grams* and *tri-grams*. We also add the length in words between two names and consider two situations: with and without length information.

The last step of the FR process is learning the model and classifying candidate pairs into FR type or *No-FR*. We apply and evaluate the designed technique

---

[3] 'Kerk van Erp' in Dutch means 'church of Erp'.

using two classifiers: a linear *Support Vector Machine* (SVM) and multinomial *Naive Bayes* [8,14] from the scikit-learn python tool[4]. In our previous work in [5] we investigated a number of other predictive models such as *Ridge regression, Perceptron, Passive Aggressive, Stochastic gradient descent, Nearest centroid* applied for document classification tasks [16]. The results are available on the web[5]. We choose SVM and Naive Bayes classifiers because SVM classifier showed the highest performance results in our previous study and Naive Bayes classifier is typically considered as a baseline in text classification [1].

Fig. 1. The illustration of tagging words in a notary act, name extraction and creation of a feature vector for a candidate pair

5 Experiments

The extraction of family relationships from notary acts and its evaluation require additional steps. The first step is the process of gathering expert opinions. This is a crucial requirement for the evaluation and training a prediction model. Therefore in this section first we present an interactive web-based interface which was used for getting input from humans. Then we elaborate on the application and the evaluation of the model. We apply 10-fold cross-validation to evaluate our method.

5.1 Notary Act Annotation

Figure 2 presents a screenshot of a developed web application for indexing family relationships. We asked experts to index notary acts and manually extract family relationships such as *parent of, siblings, married to, widow of, etc.*. Experts perform pairwise data annotation via the interface. First, they identify two person names that have a relationship then they specify the relationship type. Using the developed tool the experts manually annotated *1,005* family relationships that belong to *347* annotated notary acts. The distribution between the different types of family relationship is provided in Table 2. It is very costly to obtain labeled data. Therefore, we need a technique which is able to learn a model using a minimal number of training examples.

---

[4] http://scikit-learn.org/.
[5] http://wwwis.win.tue.nl/amontes/ecir2015/results.html.

Table 2. Statistics of manual annotation

	Marriage	Parent-child	Widow of	Sibling to	Nephew of
Number of relationships	530	298	121	45	11
Number of different relationship descriptors	43	35	21	17	4

Notary act

Theunis Jacobs en Johanna Laaracker, e.l. hebben verkocht aan Jan Lom en Gertruijd Peters, e.l. en hun erven : een stuk bouwland groot ca. 2 kleine morgen gelegen onder St.Agatha, ressort de Hoofdbank van Cuijk, jaarlijks belast met 3 malder en 1 schepel roggepacht en 2 koppels of 4 hoenders thijns beide a/d Heer van Overschie , verder vrij allodiaal erf uitgezonderd het contingent in de gemeente lasten en schattingen en met zodanige actieve en passieve servituten als tot dit perceel bouwland behoren. Het recht van de 40e pennings is aan W.G.van Oijen betaald.

Person 1   Relationship   Person 2

is married to   .

Add relationship

Relationships in this document

• Theunis Jacobs is married to Johanna Laaracker Delete

• Jan Lom is married to Gertruijd Peters Delete

Names without relationships in this document

• W.G.van Oijen Delete

Next act

Fig. 2. The designed web-based interface for annotating person names and family relationships

5.2   Result Evaluation

We evaluate the performance of the applied algorithms in terms of precision, recall, and F-score. Figure 3 show the results for different feature configurations and two classifiers: a Support Vector Machine (Fig. 3a–d) and Naive Bayes (Fig. 3e–h). The maximum *f-score* we achieve for marriage relationships using the SVM classifier and a combination of bi-grams, tri-grams of words and length between two names as presented in Fig. 3. Marriage relationships are the most frequent among other FR types, and as such more training examples are available. Another reason is that *marriage* relationship is explicit and it is clearly mentioned in the text, in contrast to *parent-child* and *siblings*. The last two types might require an additional propagation. For instance, if a mother and her two kids are mentioned in the text, then these two children are siblings of each other. In this case we first need to predict correctly parent-child links and then retrieve sibling relationship for parents that have more than one kid. The relationships *widow of* are also explicit relationships and the classifier recognizes them with the *f-score* above 0.4 despite of very small number of training examples. Overall, the SVM classifier outperforms Naive Bayes. We see that combining features together improves the SVM classification.

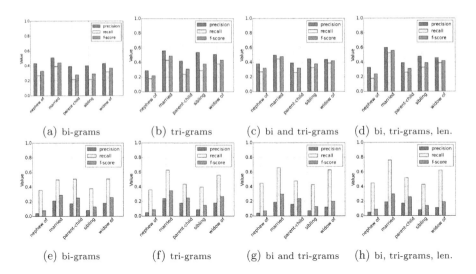

Fig. 3. Comparison of performance results for different feature configurations: (a)–(d) after applying the SVM classifier, (e)–(h) after applying Naive Bayes classifier. Len. stands for length (Color figure online)

6   Conclusions

In this paper we introduced a framework for family relationship extraction from historical notary acts. We examined different feature combinations and presented the initial results produced by two machine learning classifiers. The performance varies for different types of relationships; however we identified many family relationships correctly. We missed some family relationships because not all of them are explicitly mentioned in the text, especially concerning parent-child and siblings relationships. As a future work we plan to retrieve an implicit family relationship that require initial prediction. Another extension concerns the predictive model, where we plan to explore the use of Hidden Markov Models [4] which is widely for the text annotation purposes. However text annotation task is very different from the task of family extraction and require a number of steps in order to convert the annotated corpora into the pair of names with a specified relationship.

Acknowledgements. Mining Social Structures from Genealogical Data (project no. 640.005.003) project, part of the CATCH program funded by the Netherlands Organization for Scientific Research (NWO).

References

1. Aggarwal, C.C., Zhai, C.X.: A survey of text classification algorithms. In: Aggarwal, C.C., Zhai, C.X. (eds.) Mining Text Data, pp. 163–222. Springer, Heidelberg (2012)
2. Bird, S., Klein, E., Loper, E.: Natural Language Processing with Python, 1st edn. O'Reilly Media Inc., Sebastopol (2009)
3. Collovini, S., Pugens, L., Vanin, A.A., Vieira, R.: Extraction of relation descriptors for Portuguese using conditional random fields. In: Bazzan, A.L.C., Pichara, K. (eds.) IBERAMIA 2014. LNCS, vol. 8864, pp. 108–119. Springer, Heidelberg (2014)
4. Eddy, S.R.: What is a hidden markov model? Nat. Biotech. **22**(10), 1315–1316 (2004)
5. Efremova, J., Montes García, A., Calders, T.: Classification of historical notary acts with noisy labels. In: Hanbury, A., Kazai, G., Rauber, A., Fuhr, N. (eds.) ECIR 2015. LNCS, vol. 9022, pp. 49–54. Springer, Heidelberg (2015)
6. Efremova, J., Ranjbar-Sahraei, B., Oliehoek, F.A., Calders, T., Tuyls, K.: An interactive, web-based tool for genealogical entity resolution. In: 25th Benelux Conference on Artificial Intelligence (BNAIC 2013), The Netherlands (2013)
7. Efremova, J., Ranjbar-Sahraei, B., Oliehoek, F.A., Calders, T., Tuyls, K.: A baseline method for genealogical entity resolution. In: Proceedings of the Workshop on Population Reconstruction, Organized in the Framework of the LINKS Project (2014)
8. Frank, E., Bouckaert, R.R.: Naive Bayes for text classification with unbalanced classes. In: Fürnkranz, J., Scheffer, T., Spiliopoulou, M. (eds.) PKDD 2006. LNCS (LNAI), vol. 4213, pp. 503–510. Springer, Heidelberg (2006)
9. Ikonomakis, M., Kotsiantis, S., Tampakas, V.: Text classification using machine learning techniques. WSEAS Trans. Comput. **4**, 966–974 (2005)
10. Kivimäki, I., Panchenko, A., Dessy, A., Verdegem, D., Francq, P., Fairon, C., Bersini, H., Saerens, M.: A graph-based approach to skill extraction from text (2013)
11. Kokkinakis, D., Malm, M.: Character profiling in 19th century fiction (2011)
12. Makazhanov, A., Barbosa, D., Kondrak, G.: Extracting family relationship networks from novels(2014). CoRR, arXiv:1405.0603
13. Mintz, M., Bills, S., Snow, R., Jurafsky, D.: Distant supervision for relation extraction without labeled data. In: Proceedings of the Joint Conference of the 47th Annual Meeting of the ACL, the 4th International Joint Conference on Natural Language Processing of the AFNLP, ACL 2009, vol. 2, pp. 1003–1011, USA, 2009. Association for Computational Linguistics (2011)
14. Sammut, C., Webb, G.I.: Encyclopedia of Machine Learning. Springer, Heidelberg (2010)
15. Santos, D., Mamede, N., Baptista, J.: Extraction of family relations between entities. In: INForum 2010: - II Simpósio de Informática (2010)
16. Sebastiani, F.: Machine learning in automated text categorization. ACM Comput. Surv. **34**(1), 1–47 (2002)

Exploiting Semantic Annotation of Content with Linked Open Data (LoD) to Improve Searching Performance in Web Repositories of Multi-disciplinary Research Data

Arshad Khan[1]([✉]), Thanassis Tiropanis[1], and David Martin[2]

[1] ECS, University of Southampton, Highfield, Southampton, UK
{aaklvll, tt2}@ecs.soton.ac.uk
[2] Geography and Environment, University of Southampton, Southampton, UK
d.j.martin@soton.ac.uk

Abstract. Searching for relevant information in multi-disciplinary repositories of scientific research data is becoming a challenge for research communities such as the Social Sciences. Researchers use the available keywords-based online search, which often fall short of meeting the desired search results given the known issues of content heterogeneity, volume of data and terminological obsolescence. This leads to a number of problems including insufficient content exposure, unsatisfied researchers and above all dwindling confidence in such repositories of invaluable knowledge. In this paper, we explore the appropriateness of alternative searching based on Linked Open Data (LoD)-based semantic annotation and indexing in online repositories such as the ReStore repository (ReStore repository is an online service hosting and maintaining web resources containing data about multidisciplinary research in Social Sciences. Available at http://www.restore.ac.uk.). We explore websites content annotations using LoD to generate contemporary semantic annotations. We investigate if we can improve accuracy and relevance in search results affected by concepts and terms obsolescence in repositories of scientific content.

1 Introduction

Current searching techniques in web repositories[1] are predominantly based on keyword instances which are matched against content paying almost no attention to analyzing semantic meaning, types of content, context and relationship of keywords and phrases. Users of such web repositories have to rely on the incidental mention of the keywords and phrases in web pages, which is a challenge for users due to information overload of today's digital age. This issue is further complicated when we look at it across various disciplines where change in language terminology and concepts might change the meanings of today's web resources and other web-based content.

---

[1] A web repository stores and provides long term online access to a collection of web sites or web resources (containing static and dynamic web pages), research papers, presentations, experimental code scripts, reports etc. funded by UK research councils. Examples include http://www.data-archive.ac.uk/, http://www.timescapes.leeds.ac.uk/, http://www.icpsr.umich.edu/icpsrweb/ICPSR/.

© Springer International Publishing Switzerland 2016
P. Braslavski et al. (Eds.): RuSSIR 2015, CCIS 573, pp. 130–145, 2016.
DOI: 10.1007/978-3-319-41718-9_7

According to [1], search engines have experienced impressive enhancements in the last decade but information searching still relies on keywords-based searching which falls short of meeting users' needs due to insufficient content meaning. Similarly [2] terms the basic Web search as inadequate when it comes to finding contextually relevant information in web archives or collection of web sites like the ReStore[2] repository. Relationship between content must be an essential component to search results retrieval in such repositories but it is often missing due to the full-text keywords-based searching.

Figure 1 shows a typical web resource development and archival process involving funding bodies e.g. UK Research Councils, multi-disciplinary teams of researchers, higher education institutions and publication of research outputs in a dedicated online space either provided by the hosting institution or websites hosting company. The users of such research resources (according to our website survey in 2011 and 2013) are predominantly research students and fellows, academics, industry professionals and even funding bodies. Figure 1 outlines the entire process starting from a funding body funding a project in a higher education institution; particular research groups work on the project (typically for 3–6 years) and publish research outputs (Web resource in the form of a website) usually on hosting institution's website. These web resources end up later on in a Web repository to sustain its content for long term online access.

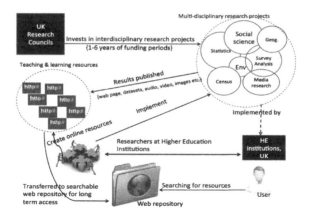

Fig. 1. An overview of web resource, creation, development and its archival into a web resource repository e.g. ReStore repository [3]

The inability to designate unambiguously the rapidly growing number of new concepts generated by the growth of knowledge and research in Social Sciences [4] is another issue failing the traditional search engines. Such issues have partly been addressed by keywords based searching where plain keyword queries are converted into equivalent semantic queries followed by syntactic normalization, word sense disambiguation [5] and noise reduction. To do that, the use of dictionaries (e.g. Wordnet), thesaurus and other library classification systems have been exploited in collaboration

---

[2] ReStore is an online repository of web resources developed as part of Economic & Social Research (ESRC) council funding-available at http://www.restore.ac.uk.

with the domain specific ontology to express keywords in more structured language. The semantic keywords are then matched with ontology terms and various semantic agents are applied to disambiguate terms and words before retrieving the results [6].

However, as described above, like other information domains, in scientific research disciplines terms change with the passage of time due to various factors e.g. cultural, social, technological, scientific and socio-economic etc. which compromise accuracy in search results. All of this suggests that semantic expressions and matching terms with ontologies classes/properties (linguistic) and instance data (semantic information) (in unstructured and heterogeneous content of repository websites) will not survive for too long and would need frequent and regular human intervention.

To address these issues, we have been focusing on 3 main areas as part of this research to see if we can improve the performance of online search applications widely used by users, who in our case, are researchers of a particular field such as the Social Sciences. These areas include (1) whether obsolescence in terms and concepts in online repositories of social science could be addressed by aiding keywords index with semantic annotation for better searching?, (2) whether a shift from domain-specific ontology-based annotation to distributed and wider data spaces like linked open-data (LoD)–based semantic annotation could address the issue of entity, concepts and relations disambiguation and thus result in better search results (high precision without compromising recall) and (3) whether crowd tagging (assigning semantic tags to relevant search results) could be employed to address the issue of content heterogeneity and obsolescence thereby benefiting the research community. This paper has so far investigated the first two points i.e. 1, 2 and thus our findings will only focus on the first two points in the rest of this paper. We will cover point 3 in the next phase of experimentation and evaluation.

This paper contributes to the related research by exploring LoD-based semantic annotation of heterogeneous websites content (ReStore repository in this case); scalability of annotation, indexing and retrieval when dealing with several content types; named entity-based linking of content based on semantic expressivity of users' keywords; and evaluating search results retrieved from the resultant semantic index, based on users' approval tagging (of system generated NE tags) and individual result ranking. Our approach further entails (a) development of an annotation, indexing and searching framework for contemporary searching in web repositories of scientific data; (b) automatic and manual annotation of content in the ReStore repository; (c) the deployment of a purpose built search engine called multi-faceted SocSci Search engine[3] for evaluating search results. We also discuss the technical deployment for various annotations and indexing approaches in the context of ReStore and NCRM[4] repositories and consider appropriate evaluation benchmarks at the time of searching and evaluation.

To evaluate search results, we have developed a custom-built search application sitting on top of full-text content, topical keywords, concepts and entities extracted from the selected documents corpus. We have conducted two distinct experiments i.e.

---

[3] Elasticsearch is a flexible and powerful open source, distributed, real-time search and analytics engine. Available at http://www.elasticsearchorg.

[4] http://www.ncrm.ac.uk.

searching performed by expert evaluators using full-text (keywords, content, title) searching technique and expert evaluation based on semantic indexing. In Sect. 2, we will review relevant work done in this area. In Sect. 3, we will elaborate the process of content annotation in repositories of heterogeneous content. Section 4 will explain the implementation of the indexing and retrieval framework and Sect. 5 will present evaluation results and lesson learnt. Section 6 will detail future work and conclusion.

2 Related Work

We recognize that some substantial work has already been done where the emphasis has been on collecting; storing and maintaining individual web resources in multi-disciplinary web repositories. However, searching across research repositories remains an open challenge. [1] highlights the limitations of keywords-based models and proposes Ontology-based information retrieval by capitalizing on Semantic Web (SW). However, poor usability of the systems usable by potential users and in completeness when applying search to heterogeneous sources of data still remain an issue.

The primary goal of any searching or retrieval system is to structure information so that it is useful to people while they search for information effectively and efficiently. Ontology-based semantic metadata extraction and storage have been around [7] since more than a decade but given that designing and evolving domain-specific ontologies still remains a challenge, alternative approaches have been adopted to extract relevant and meaningful information from text. For instance semantic indexing and retrieving the resulting knowledge base in a scalable manner still remains an issue. The semantic web research community has been experimenting with a fixed set of documents corpora with non-user friendly web based interfaces, which limit users' browsing capacity to visualize search results thus affecting overall information utility. Another major problem, the semantic web community faces for the construction of innovative and knowledge-based web services is to reduce the programming effort while keeping the web preparation task as small as possible [8].

We have seen user's query expansion-based searching and ontology-based information retrieval model proposed by [1, 9] and ontology-extension model based on adding further classes to the root ontology in [10] and ontology classes/properties matching between LoD cloud datasets (DBPedia, Freebase, Factforge etc.) and domain independent ontology like PROTON [11]. However, the level of complexity and the amount of time, it takes to refine the classes and their relationship with external sources of data (e.g. concepts disambiguation, over linking, word sense and terms stemming), leaves web scalability as a non-addressed issue. All such approaches tend to distort the actual users queries [12] thus turning the words in ambiguous queries leading to less relevant and imprecise search results. Another reason for ruling out extending a general purpose light-weight ontology like PROTON [13] to include new classes and properties in a domain ontology along with defining and applying subsumption rules, is the lack of experts and participants (to continuously evaluate new classes) in a scientific field such as Social Sciences. Moreover continuously monitoring the emergence of new concepts and terminologies in a particular domain specific ontology like Social Science Research Methods is in itself unsustainable and prone to ambiguities at every stage.

On the another hand, [14] proposes key-phrase extraction based on semantic blocks which entails pre-selecting blocks of information that have higher coherence in terms of extracting the most meaningful key terms from a web page. Such an approach further complicates the already presumptuous approach of ontology-based entity and concepts extraction by adding another assumption that a more coherent space in a web page or web documents will be preselected before annotating the content inside.

Ontology availability, development and evolution make it a hard choice for developers who are mainly responsible for the implementation of semantic search web applications. We recognize that KIM (Knowledge & Information Management) offers a running platform for ontology-based annotation and retrieval [15] but amalgamating the built in KIM ontology with domain specific ontology followed by gazetteer-based annotations require huge efforts both on the part of developers and ontology designers. KIM server-based search application is still far from being implemented in typical client server architecture, which is what we have experimented with as part of this research.

3   Our Methodology

We describe a framework that incorporates semantic text analysis of content in ReStore repository to add semantic metadata and topical keywords to build a keywords and semantic Knowledge Base (KB). To address the issue of ambiguous terms during annotation, extend the semantic meaning of concepts in web pages and sustain the meaning of terms and concepts in scientific repositories with the passage of time, we have adopted Linked Open Data (LoD) as a tagging data source for various types of documents in our repository. The LoD numbers over 200 datasets which span numerous domains such as media, geography, publications and life sciences etc. incorporating several cross-domain datasets [16]. It is an open source of structured data, which so far has been employed for building Linked Data (LD) browsers, LD search engines and LD domain-specific application such as semantic tagging [17]. A number of web services have been developed recently to extract structured information from text (incorporating LoD) such as Alchemy API[5], DBPedia Spotlight[6], Extractive[7], OpenCalais[8] and Zemanta[9]. We have used Alchemy API due to its holistic approach towards text analysis and broad-based training set (250 times larger than Wikipedia) used to model a domain like ours. The tool uses machine learning and natural language parsing algorithms for analyzing web or text-based content for named entity extraction, sense tagging and relationship identification [18]. Alchemy API was also one of the best in the performance evaluation review of [19] where Alchemy API remained the primary option for NE recognition and overall precision and recall of NEs and types inferences and URI disambiguation.

---

[5] http://www.alchemyapi.com.

[6] http://dbpedia.org/spotlight.

[7] http://extractive.com.

[8] http://www.opencalais.com.

[9] http://www.zemanta.com.

We start by making the case for information retrieval from the KB and build on this by evaluating search results in terms of TREC[10] 11-points Precision/Recall measures to assess accuracy in search results. We also measure users' tanking to assess the level of users' satisfaction (degree of relevance), which extends the meaning of relevance from binary (relevant or non-relevant) evaluation to users' satisfaction evaluation. We describe the framework as comprising of four elements i.e. (a) Semantic structuring of ReStore repository's content (Schematization); (b) Mass annotation of content addressing all types of content (static/dynamic); (c) Indexing semantic annotation along with actual content (Semantic annotation for metadata augmentation); and (d) Web-based search application using Elasticsearch distributed searching (for precision/ relevance evaluation in search results).

3.1   Annotation of Heterogeneous Content with LoD

As it can be seen in Fig. 2 below, we have setup an environment where content from two web sites i.e. www.restore.ac.uk and www.ncrm.ac.uk are extracted by using two distinct methods i.e. crawling static web pages from the websites and extracting dynamic content and non-web page documents from database management systems e.g. MySQL. Figure 2 shows our four framework components i.e. schematization, annotation, indexing, and retrieval in action. Semantic annotation of web documents is performed using one of the best semantic annotators i.e. Alchemy API which analyze each document by using built in NLP (Natural Language processing) and Machine Learning (ML) and other complex linguistic, statistical, and neural network algorithms.

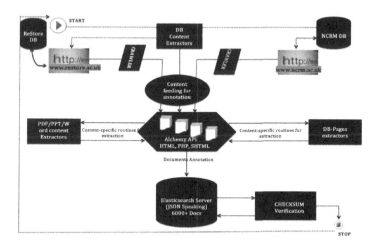

Fig. 2. Web repositories extraction, annotation and indexing process flow

This service crawls billions of pages every month thus expanding its knowledge base through which entities and concepts are identified in web documents and linked to

---

[10] Text Retrieval Conference http://trec.nist.gov/.

various linked data sources of data. We obtained special license from the Denver-based company allowing us to analyze 30,000 documents per day. Similarly we have added topical keywords, concepts and entities to 3000+ documents and have stored the relevant TF.IDF score along with Alchemy API score against each of the individual items in a single record to enable ourselves to manipulate precision and recall during various experiments and evaluation exercises in later sections. We have indexed data in multiple indexes based on the type of data, size of documents and the possibilities in which the indexed data could be searched and browsed. We have achieved this by designing unique schemas for each index. Finally, we have developed a searching application, which sit on top of the above to facilitate users search for their topic of interest as part of evaluation exercises.

3.2   Scope of Annotation and Indexing

The scale, at which we have setup the annotation, indexing and search results evaluation environment, is extensible (for example beyond ReStore content) and offer greater degree of freedom in terms of utilizing annotation APIs e.g. Concepts, Entities and Keywords. We can for instance use sentiment analysis in order to determine the degree of relevance of concepts, entities and keywords in our documents across different LoD datasets. That in essence offers various search performance levels which could be evaluated at the time of searching to attain the required level of performance. We have for example evaluated the relevant concepts, tagged by our system, by presenting most relevant tags to evaluators next to each search results link as detailed in the evaluation section of this paper. Unlike evaluating Ontology classes of a specific domain, assessing the appropriateness of a sub class in a domain Ontology before annotating content with it, we are interested in annotating topic of interests in a range of topically diverse heterogeneous content contained in web pages, portable PDF documents, presentations and other file formats. To assess the appropriateness of annotation, we present search results to evaluators based on topic (term, concept, keywords) popularity and weight of relevance calculated at the time of annotating and indexing.

4   Implementation

To start off, we have subdivided the annotation process into two distinct categories i.e. 1. Annotation of content based on topic keywords using Alchemy Keywords API and 2. Semantic annotation of content using Alachemy Concepts, Entities APIs. The content includes web pages, PDF, CSV, Word, Powerpoint presentations and other software code script produced as part of various experiments by the researchers during the course of research projects.

4.1   The Annotation Process

Figure 3 shows the entire annotation process along with web-based search interface for evaluation purposes. The process flow starts from the diamond-shaped box which is

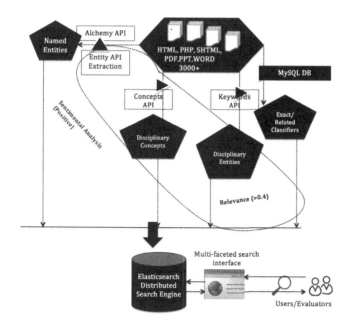

Fig. 3. Automatic annotation of content extracting keywords, entities and concepts using Alchemy APIs (incorporating LoD datasets i.e. DBPedia, YAGO [20], OpenCyc (OpenCyc contains hundreds of thousands of general knowledge terms organized in a carefully designed ontology. Available at http://opencyc.org/doc/opencycapi) and Freebase [21]) keywords entity and concepts extraction

our benchmark document corpus hosted by two LIVE websites i.e. NCRM and ReStore.

By using Alchemy API service, we take the entire corpus of structured documents as a Knowledge Base which conform to the data retrieval model elaborated by [1]. With semantic annotation two important tasks of the semantic web can be achieved i.e. (1) extracting and hyperlinking named entities in documents and (2) finding relevant documents in accordance with entities [22]. All this brings structure to web content in order to enhance the meanings of text in web pages, which improve content searching based on mutual relationships between different sections of documents, keywords and entities. Based on this interpretation, we assume that (a) we have an entity or context extraction platform which would be applied on a number of documents in order to pinpoint keywords, entities and concepts to a more meaningful contemporary source of data e.g. DBPedia, Freebase and Yago; and (b) build a Knowledge Base of data which comprise actual documents and semantic metadata along with inter-documents relationships.

4.2   Elasticsearch Search Engine as a Knowledge Management Platform

We have used Elasticsearch server mounted on the current ReStore web server with an intention to turn it into a full-fledged dedicated semantic and full text search server used with the front end (PHP, JavaScript, Elastica Library) to display search results to users for evaluation purposes. We have addressed a particular issue which many semantic search system suffers from i.e. usability limitations where users are expected to use formal query language to express their requirements and lack of optimal semantic annotation of content in web documents emanating from using small set of pre-defined domain ontologies and datasets [1]. Using DBPedia spotlight discussed in [23], for instance assumes that users should be able to opt for preferred or alternative labels while searching for things in the DBPedia spotlight web application. We understand, that such assumptions compromises the soundness of semantic data-based web application as the majority of users still prefer to use free text keywords based search without pre-specifying advanced search options [9].

Elasticsearch analyzers first analyze all the content belonging to each document via JSON-formatted URLs and relevant scores are stored against keywords, entities and concepts (extracted by Alchemy API using three different API services i.e. Keywords, Entity and Concepts). Each document D_j represents a vector space model in the following manner:

$$D_j = (t_k, t_e, t_c \ldots, t_{kec})$$

Where t_k, t_e, t_c, t_t are the keywords (k), entities (e) and concepts (c) terms. With this representation, each document is a vector having the above elements for influencing ranking of search results. The scoring algorithms are based upon statistical and NLP techniques employed by Elasticsearch distributed search server. It is however to be mentioned here that we have indexed individual Alchemy APIs-based score as well in each index in order to run sub queries based on users' browsing preferences but browsing based evaluation is beyond the remit of this paper.

5   Evaluation

We have analyzed the performance in terms of search results relevance, precision and recall in two different categories i.e. (1) searching on the basis of keywords and actual content (2) searching on the basis of topic keywords, semantic concepts and entities extracted by Alchemy Keywords, Concepts and Entities APIs. In our benchmark document collection, we have annotated more than 3000 documents and have built a semantic store ready to be exploited by our web-based search application. The whole process is carried out in a client-server architecture which includes ReStore web server, ReStore and NCRM Database server containing 5 different MySQL Databases populating almost 6000 web pages in NCRM website and 3000 in ReStore.

Similarly in our queries benchmark collection, we have captured a set of free-text queries from Google Analytics which were submitted by online users of the ReStore website to find the desired information using the current web search. The total number of queries randomly selected was 34. The criteria for selecting these queries included the number of clicks they generated and brought users to the ReStore repository site, recency of use and meaningfulness of terms in the query. However, to reduce bias in all 34 queries, we selected queries having one single term, multiple terms and mixture of keywords and concepts. Here is a sample of user queries in the benchmark query collection. {*cohort sequential design, design effects in statistics, paradata in survey research, randomized control trials, evaluating interaction effects, ethnic group, Forecasting, Stages of a systematic review, sample enumeration, what is media analysis.*}

With regard to the experts' judgment evaluation, our evaluators included Librarians, Social Science academics, Social Science research fellows and PhD students coming from various disciplines e.g. Social Sciences, Education, Geography & Environment and Statistics. Feedback was collected from 15 expert evaluators over a period of 3 weeks. The evaluation exercises were designed in such a way that they had a freedom in reviewing 70 web documents at their own pace as long as they were logged in. We asked the evaluators to carry out search exercises by using the pre-selected set of queries. Their assessment included whether (a) a search result is relevant or not after viewing the content of the results by clicking on the link; and (b) ranking the result in terms of the number of stars corresponding to each results; and (c) authenticating potential concepts/entities from the list having association with each result. The first 10 results retrieved were assumed to be relevant against each query. Each expert user was given a set of 7 queries before logging on to the search results evaluation page and his/her activity was recorded in the database along with their ranking of each search results. It is to be mentioned here that some queries were evaluated by only two evaluators in which case we averaged the ranking before using it in our evaluation analysis. On the basis of their assessment, we have computed MAP (Mean Average Precision) and have drawn TREC's (Text Retrieval Conference)[12] points recall precision curves in the next section to show that enhanced semantic metadata attachment to the actual content clearly improves precision in search results with maximum recall. We have also computed ranking performed by users (MAR-Mean Average Ranking), which reflects the system's ranking in terms of accuracy and results relevance.

5.1   Search Results Evaluation

We expected each participant to evaluate 70 search results in total i.e. 10 against each query (7 queries per participants). All together 15 expert evaluators evaluated the system and 886 web documents were evaluated. These participants also added 2555 semantic concepts and entity tags with these 886 web documents. Our system presents a list of 10 results to users with a summary for each highlighting matched words in the query, which helps users make a quick sense of the result before clicking on it. We assume that typically every web users would want every result on the first page to be relevant (high precision) but have little interest in knowing let alone looking at every document that is relevant. We have used precision/recall measures to determine the system performance. Recall is the ratio of relevant results returned divided by all relevant results and precision is the ratio of the number of relevant records retrieved to the total number of irrelevant and relevant results retrieved.

While calculating query-level precision, recall and Average Precision (AP) we assumed all top 10 documents retrieved against each individual query to be relevant. We calculated Average Precision (AP) on query level using TREC's 11-points recall and compared it with that of keywords-based AP to ascertain which system performed better in terms of precise search results without compromising too much recall. To properly quantify the level of relevance in both scenarios, we have used the combined measures that assess the precision/recall tradeoffs which is given by:

$MAP = (\sum_{i=1}^{Q} AP_i)/Q$ where Q = number of queries in a batch. We have also used MAR (Mean Average Ranking) to ascertain the degree of relevance in terms of users' happiness based on users' ranking.

5.2 Semantic Entities and Concepts Tagging

Alongside actual results evaluation and ranking of each result, participants also tagged relevant concepts and entities, presented to them by the system next to each result. The tagging has been of help in understanding participants' decision element for ranking a particular result. For example when "forecasting" is searched, one of the concepts that was suggested to users for tagging in a few results was "prediction", "decision theory", "Bayesian inference" and "statistical inference". Those relevant concepts had already been identified by the semantic annotator but participants' tagging enabled us to re-validate the system's accuracy, which is reflected in assessing the degree of user happiness or MAR in Fig. 6. In contrast, when forecasting was searched in the existing online search facility of ReStore, most of the results in top 10 results, were retrieved because of the mention of the word forecasting. Likewise, when forecasting was searched by multiple users as part of our evaluation, they highly ranked a result which was no 3 in the top 10 list and it had no mention of forecasting but the content were about a research tool used to predict housing, income and education situations of participants taking part in a case study. Google's top 10 results included those defining forecasting, Meteorological office forecasting and baseball game forecasting.

The better performance in semantic search results evaluation in Figs. 4 and 5 is attributed to the semantic index scoring criteria. For example, one of the queries *mixture model"* doesn't exist in the document vector space under the "keywords" list but it has a high score under the "concepts" list of that document. Similarly another query *"Randomized control trials"* exists in the document space under "keywords" list but with a low score and high score under the "concepts" list. Thus when searched, the document having such concept (not necessarily under keywords) with high score was presented to user as highly relevant. Likewise, when *"reasoning"* was searched, a document containing *"critical thinking"* concept came first in top 10 search results which was tagged as relevant by the evaluators.

5.3 The Experiments

After entering individual queries in the search box, a participant was expected to classify a result as either relevant or irrelevant i.e. 1 for relevant and 0 for irrelevant web documents. The participant also star-rated the result and tagged relevant concepts and entities retrieved along with individual results, which can be used for measuring average ranking across the set of queries.

For instance we have to see how many relevant pages $r = \{r^1, r^2, \ldots r^n\}$ could be retrieved in top 10 pages which were retrieved against each query from keywords index $Q(k) = \{k^1, k^2, k^3 \ldots k7\}$ and semantic index $Q(s) = \{s^1, s2, s3 \ldots s^7\}$. In other words, how best could our system interpret keywords in users' queries, turn them into topical keywords, concepts and entities and retrieve those web pages, annotated by the annotator during the annotation and scored at the time of indexing. Precision at k documents has been our assumption throughout the experimentation process, which implies that the best set of search results appears on the first page of results set and the total number of best results is 10. Similarly recall at k documents is based on our assumption that the relevant documents at the time of submitting each query will remain 10 documents. This approach has been adopted based on the nature of web searching in ReStore repository which host archived content and most users are by and large interested in the first 10 results to maximize their satisfaction in terms of finding relevant results.

5.4 Fixed vs. Interpolated Precision Measurement

We assume during our evaluation that a user will examine a fixed number of retrieved results and we will calculate precision at that rank and interpolated rank. Hence our fixed average precision is given by: $P_{(n)} = \sum_{n=1}^{N} \frac{r(n)}{n}$ where r(n) is the number of relevant items (at cut off k relevant document) retrieved in the top n which in our case is 10 documents (N) at each level of individual information needs in the form of user queries. However, if $(k + 1)th$ retrieved is not relevant, precision will drop but recall will remain the same. Similarly if $(k + 1)th$ document is relevant both precision/recall increase. Therefore we had to extend these measures by using ranked retrieval results, which is a standard with search engines. Interpolated precision is therefore given by:

$$P_{11-pt} = \frac{1}{11} \sum_{j=0}^{10} \frac{1}{N} \sum_{I=1}^{N} P_i(r_j)$$

where $P(r_j)$ is the precision at our recall points but $P(r_j)$ doesn't coincide with measurable data point r if number of relevant documents per query is not divisible by 10 in which case, the interpolated precision is given by:

$P_{interpolated}(r) = \max\{P_i : r_i \geq r$ where (Pi, ri) are raw values obtained against different queries or information needs. So the new average interpolated precision is given by:

$$P_{11-pt-interpolated} = \frac{1}{11} \sum_{r \in \{0, 0.1, \ldots, 1.0\}} p_{interp}(r)$$

In other words, interpolated precision shows maximum of future precision values for current recall points. The normal precision/recall curve reacts to such variations differently as we have shown in Fig. 4. An increase in both precision and recall means, the users is willing to look at more results. All this tells us about expected precision/recall values for another set of results (k + 1,2,3...n). Since $P_{(n)}$ ignores the rank position of relevant documents retrieved above cut off (i.e. 10 + 1), we have calculate interpolated precision at each query level to assess the system performance at n + 1 documents which is beyond the existing cut off point.

Figure 4 shows the TREC-11 points ranked retrieval precision/recall curve which is representative of the two systems we have assessed i.e. topical keywords or full-text search vs. semantic index-based searching.

Figure 4 shows TREC 11 points Interpolated Precision and Recall curve showing system performance i.e. keywords (full-text) searching vs. keywords & semantic index-based searching. It also shows averages system's performance over the entire queries batch.

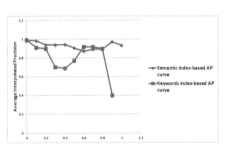   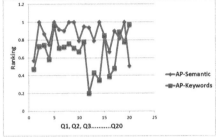

Fig. 4. TREC 11 points Interpolated P/R   **Fig. 5.** Non-interpolated AP curve

We can clearly see that the behavior of un-interpolated keyword index-based precision-recall curve is quite fluctuating while that of semantic index remains firm in interpolated Fig. 4, vacillating between 100 and 80 % for the first top 10 results for the entire queries batch. Of course we have given some ground to the fact that only 20 out of 34 queries were attempted by two different types of evaluators i.e. a batch of 7 queries was attempted by two evaluators one based on full-text search index and another semantic index. We also averaged multiple queries in order to get the overall performance of keywords based and semantic index-based search system. We also calculated non-interpolated average precision at each query level which is given by: $P(r) = \sum_{i=1}^{N_q} \frac{P_i(r)}{N_q}$ where $P(r)$ is the average precision at Recall level r and N_q is the number of queries. $P_i(r)$ is the precision at Recall level r for the i-th query. Figure 5 clearly shows better performance in terms of un-interpolated precision and recall when queries were searched against semantic index. Performance in both situations suggests that semantic index based curve performs better. Figure 5 indicates overall better performance with the exception of few queries where keywords index-based curve performs better. But this has been offset by the interpolated precision-recall curve in Fig. 4 where semantic-index-based performance remains consistent.

We have also calculated Mean Average Precision (MAP), which has become commonplace in recent years providing a single-figure measure of quality across recall levels. Using MAP, fixed recall levels are not chosen and there is no interpolation. $MAP = (\sum_{i=1}^{Q} AP_i)/Q$ where Q = number of queries in a batch. MAP ensures that equal weightage is given to all queries i.e. those containing rare and common terms with different recalls. Our MAP for keywords and semantic search results are 66 % and 84 % respectively.

5.5   Mean Average Ranking

Precision and Recall curve don't allow as such for the degree of relevancy when it comes to retrieving precise and relevant documents. This is partly because this measures is based on binary classification and individuals' perception. What is relevant to one person may not be relevant to another. To address this issue, we have averaged ranking of all relevant documents against each of all queries in our experimental batch and have shown the average ranking in the following graph to prove our point. We assume that along with measuring the degree of relevance in search results, it is also equally important to measure the utility or satisfaction level achieved by users after exploring through the actual content. Mean Average Ranking (MAR) therefore represents our ranking model, which have applied in the following diagram.

$MAR = (\sum_{j=1}^{Q} AR_j)/Q$ where Q = number of queries in a batch.

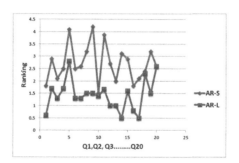

Fig. 6. Average ranking curve shows averages ranking over a medium set of queries. We have computed average ranking at each information need or query level and have concluded that MAR in semantic index-based searching performs better than the full-text index

6   Conclusion and Future Work

The aim of this research has been to explore new avenues for semantic index-based searching and assessment of users' happiness at the time of searching for relevant results. By correct identification of LoD-based topical keywords, concepts and entities, users could continue finding relevant information using online search regardless of the time factor. In other words by regularly adding semantic metadata using LoD, content will continue to sustain its value through enhanced semantic metadata annotation and indexing using online search applications regardless of the time and discipline elements. To advance this research, we will further investigate users' contribution or crowd-annotation (using controlled vocabulary and free text tags) at the time of exploring content web repositories and its impact on search results. Using built-in annotation tools in every web page of the ReStore and NCRM repositories (already deployed in benchmark documents), we will conduct focus group exercises to (a) crowd-annotate various content in the above online repositories to address concepts obsolescence; and (b) sustain the quality of search results regardless of time factor in the designated repository of social science research data.

References

1. Fernandez, M., et al.: Semantically enhanced information retrieval: an ontology-based approach. Web Semantics: Science, Services and Agents on the World Wide Web 9(4), 434–452 (2011)
2. Wu, P.H., Heok, A.K., Tamsir, I.P.: Annotating the web archives – an exploration of web archives cataloging and semantic web. In: Sugimoto, S., Hunter, J., Rauber, A., Morishima, A. (eds.) ICADL 2006. LNCS, vol. 4312, pp. 12–21. Springer, Heidelberg (2006)
3. Khan, A., Martin, D., Tiropanis, T.: Using semantic indexing to improve searching performance in web archives. In: International Journal on Advances in Internet Technology, Seville, Spain, pp. 1–4 (2012)
4. Riggs, F.W., Interconcept report: a new paradigm for solving the terminology problems of the social sciences, UNESCO, vol. 44 (1981)
5. Snow, R., et al.: Cheap and fast—but is it good? Evaluating non-expert annotations for natural language tasks. In: Proceedings of the Conference on Empirical Methods in Natural Language Processing (2008). Association for Computational Linguistics
6. Royo, J.A., et al.: Searching the web: from keywords to semantic queries. In: Third International Conference on Information Technology and Applications, ICITA 2005 (2005)
7. Zervanou, K., et al.: Enrichment and structuring of archival description metadata. In: ACL HLT 2011, p. 44 (2011)
8. Benjamins, R., et al.: The six challenges of the semantic web (2002)
9. Yang, C., Yang, K.-C., Yuan, H.-C.: Improving the search process through ontology-based adaptive semantic search. The Electronic Library 25(2), 234–248 (2007)
10. Georgiev, G., et al.: Adaptive semantic publishing. In: WaSABi@ ISWC (2013)
11. Damova, M., et al.: Mapping the central LOD ontologies to PROTON upper-level ontology. In: Proceedings of the Fifth International Workshop on Ontology Matching (2010)
12. Shabanzadeh, M., Nematbakhsh, M.A., Nematbakhsh, N.: A semantic based query expansion to search. In: 2010 International Conference on Intelligent Control and Information Processing (ICICIP) (2010)
13. Popov, B., Kiryakov, A., Kirilov, A., Manov, D., Ognyanoff, D., Goranov, M.: KIM – semantic annotation platform. In: Fensel, D., Sycara, K., Mylopoulos, J. (eds.) ISWC 2003. LNCS, vol. 2870, pp. 834–849. Springer, Heidelberg (2003)
14. De Virgilio, R.: RDFa based annotation of web pages through keyphrases extraction. In: Meersman, R., et al. (eds.) OTM 2011, Part II. LNCS, vol. 7045, pp. 644–661. Springer, Heidelberg (2011)
15. Bai, R., Wang, X.: A semantic information retrieval system based on KIM. In: 2010 International Conference on E-Health Networking, Digital Ecosystems and Technologies (EDT) (2010)
16. Rusu, D., Fortuna, B., Mladenic, D.: Automatically annotating text with linked open data. In: LDOW (2011)
17. Bizer, C., Heath, T., Berners-Lee, T.: Linked data-the story so far. In: Semantic Services, Interoperability and Web Applications: Emerging Concepts, pp. 205–227 (2009)

18. Gangemi, A.: A comparison of knowledge extraction tools for the semantic web. In: Cimiano, P., Corcho, O., Presutti, V., Hollink, L., Rudolph, S. (eds.) ESWC 2013. LNCS, vol. 7882, pp. 351–366. Springer, Heidelberg (2013)
19. Rizzo, G., et al.: NERD meets NIF: lifting NLP extraction results to the linked data cloud. In: LDOW, vol. 937 (2012)
20. Suchanek, F.M., Kasneci, G., Weikum, G.: Yago: a core of semantic knowledge. In: Proceedings of the 16th International Conference on World Wide Web, pp. 697–706. ACM, Banff, Alberta, Canada (2007)
21. Bollacker, K., et al.: Freebase: a collaboratively created graph database for structuring human knowledge. In: Proceedings of the 2008 ACM SIGMOD International Conference on Management of data, pp. 1247–1250. ACM, Vancouver, Canada (2008)
22. Kiryakov, A., et al.: Semantic annotation, indexing, and retrieval. J. Web Semant. **2**(1), 49–79 (2004). Elsevier's
23. Mendes, P.N., et al.: DBpedia spotlight: shedding light on the web of documents. In: Proceedings of the 7th International Conference on Semantic Systems, ACM (2011)

Construction of a Russian Paraphrase Corpus: Unsupervised Paraphrase Extraction

Ekaterina Pronoza[✉], Elena Yagunova, and Anton Pronoza

Saint-Petersburg State University, Saint-Petersburg, Russian Federation
katpronoza@gmail.com, iagounova.elena@gmail.com, antpro@list.ru

Abstract. This paper presents a crowdsourcing project on the creation of a publicly available corpus of sentential paraphrases for Russian. Collected from the news headlines, such corpus could be applied for information extraction and text summarization. We collect news headlines from different agencies in real-time; paraphrase candidates are extracted from the headlines using an unsupervised matrix similarity metric. We provide user-friendly online interface for crowdsourced annotation which is available at paraphraser.ru. There are 5181 annotated sentence pairs at the moment, with 4758 of them included in the corpus. The annotation process is going on and the current version of the corpus is freely available at http://paraphraser.ru.

Keywords: Russian paraphrase corpus · Lexical similarity metric · Unsupervised paraphrase extraction · Crowdsourcing

1 Introduction

Our aim is to create a publicly available Russian paraphrase corpus which could be applied for information extraction (IE), text summarization (TS) and compression. We believe that such corpus can be helpful for paraphrase identification and generation for Russian and that is why we focus on the sentential paraphrases. Indeed, a sentential corpus does not impose any specific methods of further paraphrase identification or generation on the researcher. If such corpus is representative enough, it can serve as a dataset for the experiments on the extraction of word-, phrase- and syntactic level paraphrases.

Paraphrase is restatement of a text: it conveys the same meaning in another form. Such natural language processing (NLP) tasks as paraphrase identification and generation have been shown to be helpful for IE [25], question answering [14], machine translation [7], TS [19], text simplification [29], etc. Paraphrase identification is used to detect plagiarism [6] and to remove redundancies in TS [19] and IE [25], while paraphrase generation – to expand queries in information retrieval and question answering [14] and patterns – in IE. Paraphrase generation is also useful for text normalization [28] and textual entailment recognition tasks [9].

As far as the definition of paraphrase is concerned, it generally implies that the same message is expressed in different words, but it does not prescribe which portion of text is replaced by paraphrasing. Neither does it state whether common knowledge can be

© Springer International Publishing Switzerland 2016
P. Braslavski et al. (Eds.): RuSSIR 2015, CCIS 573, pp. 146–157, 2016.
DOI: 10.1007/978-3-319-41718-9_8

used when judging on the similarity of the two messages. As a consequence of this ambiguity, some researchers believe that paraphrases should have absolute semantic equivalence while others allow for bidirectional textual entailment, when the two messages convey roughly the same meaning.

Let us consider an example from our corpus: (1) ВТБ может продать долю в Tele2 в ближайшие недели. /VTB might sell its shares in TELE2 in the nearest weeks/ (2) ВТБ анонсировал продажу Tele2. /VTB announced the sale of TELE2/

Although it is clear that the two sentences describe the same event, the first one has additional details: indication of the time and the fact that the shares are going to be sold. A human judger, with his/her knowledge about the world, might consider these sentences paraphrases. But if we intend to teach a machine to identify semantically equivalent paraphrases, a threshold for paraphrases should be higher. On the other hand, the second sentence can be considered a summarization of the first one, and therefore such types of paraphrases can be used in automatic TS.

In our research, we intend to construct paraphrase corpus for IE and TS. We believe that the former task requires semantically equivalent, or precise paraphrases while the latter one demands roughly similar ones (so-called loose paraphrases) like those in our example. Thus, it is important for us to distinguish precise paraphrases (PP) and loose paraphrases (LP) while constructing our paraphrase corpus.

Today there are already a number of available paraphrase resources, Microsoft Paraphrase Corpus being the most well-known of them [13]. A wide number of metrics for paraphrase identification (for English) are evaluated against this corpus.

For Russian there are no publicly available paraphrase resources known to us, with the only exception of the dataset published by Ganitkevich et al. as part of The Paraphrase Database project [17]. The latter includes paraphrases on the word-, phrase- and syntactic levels, and each paraphrase pair is annotated with the set of count- and probability-based features. Such corpus can be used for both IE and TS, but it lacks information on the context of paraphrases. We believe that if such context (the original sentences) was provided, it could improve both these NLP tasks. That is why we aim at constructing a sentential corpus.

Thus, our task is to construct a corpus with both PPs and LPs and to make it helpful for paraphrase identification and generation in IE, TS and text compression tasks.

Our research is a part of an ongoing crowdsourcing project available at paraphraser.ru, with our current results available at paraphraser.ru/scorer/stat.

2   Related Work

In paraphrase identification/generation, unlike many other NLP applications, the data is usually hard to get. Paraphrases do not emerge naturally, like users' clicks or query logs, and gathering them manually is a tedious task. Moreover, one usually needs large amounts of data to collect just a few paraphrases.

Paraphrase corpora can be constructed from "natural" or "artificial" sources. The former usually include parallel multilingual corpora [2] and comparable monolingual corpora [22] (different translations of the same texts [11]; news texts/clusters [1, 10, 13,

27]; texts on similar topics, e.g., from the social networks (e.g., Twitter Paraphrase Corpus) [28] or students' answers to the questions [9]; social media [3], Wikipedia [26]; different descriptions of the same videos [8]), etc. "Artificial" sources are texts paraphrased by humans [20, 21, 24].

Gold standard paraphrases are typically extracted from the candidates set using either experts' [13, 20] or crowdsourced [1, 6, 8] annotation. 2-way and 3-way annotation is a common approach, but sometimes a complex system of characteristics is introduced (e.g., in [20] paraphrases are annotated along 10 dimensions of paraphrase characteristics on a 6 point scale).

A detailed overview of all the existing paraphrase corpora is beyond the scope of this paper. A thorough and insightful review of different sentential paraphrase datasets can be found in [21] where the authors present recommendations on paraphrase corpora construction and raise a number of important problems to the community.

Due to the aim of our research and space limitations we inevitably focus on the well-known Microsoft Research Paraphrase Corpus (MSRP) and on The Paraphrase Database, the only publicly available resource of Russian paraphrases known to us.

MSRP is not the oldest paraphrase corpus, but is definitely the one which greatly inspired research in paraphrase community. It was constructed as a broad-domain corpus of sentential paraphrases which would be amenable to statistical machine translation (SMT) techniques [13]. It consists of 5801 pairs of English sentences collected from news clusters and annotated by 2 experts. An initial set of paraphrases is extracted using Levenshtein edit distance. The authors only consider first 3 sentences of the articles and apply several criteria to their length and lexical distance between the sentences. The resulting dataset is extracted using SVM with morphological, lexical, string similarity and composite features.

Although MSRP is widely used as the gold standard in the experiments on paraphrase extraction methods, it is often criticized by researchers for its loose definition of paraphrase, for its 2-way annotation, high lexical overlap, etc.

While constructing a Russian corpus, we try to solve the problem of paraphrase ambiguity by distinguishing 2 types of paraphrases: precise and loose ones. We have 3-way annotation: precise paraphrases, loose paraphrases and non-paraphrases. As for the lexical overlap problem, we consider this overlap acceptable and even helpful in our case. Russian is a language with free word order, and pairs of sentences which consist of the same words put in different order could be used for learning syntactic patterns for paraphrase generation.

As we have already mentioned, there is one publicly available Russian paraphrase resource known to us: the dataset published by Ganitkevich et al. as part of The Paraphrase Database project (PPDB) [17]. The authors collected an impressively large database of paraphrases on word-, phrase- and syntactic levels. Syntactic level paraphrases are annotated with nonterminal symbols (constituents, in terms of phrase structure grammar) and contain placeholders which can be substituted with any paraphrase matching its syntactic type. In addition, all types of paraphrases are annotated with count and probability-based features. These features include the difference in the number of words/characters/average word length between the original phrase and the paraphrase, the probability of the original phrase given the paraphrase, alignment features, etc.

Some features are derived from the syntactic rules, e.g., the probability of the lefthand side nonterminal symbol given the paraphrase (and vice versa).

The training data for Russian is substantial in PPDB (over 2 million sentence pairs), and the resulting dataset is large as well. It is collected from the corpora typically used in SMT: CommonCrawl, Yandex 1M corpus and News Commentary. The authors use a language independent method to extract paraphrases from parallel bilingual texts: paraphrases are found in a single language by "pivoting" over a shared translation in another language. Such approach was introduced by Bannard and Callison-Burch in 2005 [2] and since then it has been successfully applied by many researchers. The authors acknowledge that in morphologically rich languages different forms of the same word tend to group into the same paraphrase clusters because English phrases are chosen as the pivot ones (in Russian different forms of the same word are considered paraphrases in PPDB). While for some tasks it could be desirable, for others it is definitely not (it could cause generation of incorrect paraphrases). Moreover, such grouping leads to the rapid growth of the dataset, and, with a number of available morphological parsers today, it seems unnecessary. We also believe that we should use language-specific methods (in contrast with language-independent ones) when dealing with a morphologically rich language.

Unlike other paraphrase resources, our corpus is not intended to be a general-purpose one. According to our tasks (IE and TS) we collect it from the news texts. The corpus consists of sentential paraphrases, and lower level paraphrase pairs can be extracted from it using any of the existing methods (e.g., SMT methods).

3   Unsupervised Paraphrase Extraction

3.1   Data: Method

We adopt a sentence-level approach and extract paraphrases from the news articles published on the Web. The latter is a truly rich source of paraphrases: the articles describing the same events appear in different newspapers every day.

Due to the lack of training data for Russian, our approach is unsupervised. It extends the one described by Fernando and Stevenson [15]: their method yields the best results against MSRP among the latest unsupervised approaches. Our hypothesis is that paraphrases can be successfully extracted from the Russian news texts based on a lexical similarity metric.

We automatically extract articles published by several newspapers on the same day during the last 2 years. Then we adopt the strategy by Wubben et al. [27] and proceed with pairwise comparisons of headlines. A headline of an article can be considered its compression[1], and we suppose that the headlines of the articles describing the same events are similar and may even be paraphrases of each other. Moreover, headlines

---

[1] This statement can only be applied to the informative news texts (the ones intended to inform, and not to persuade the reader) and not to the publicistic texts (exerting influence on the reader in the first place). A publicistic headline is often designed to attract readers' attention. However, both publicistic and informative texts can be used as a source of paraphrases.

comparison is much faster than the comparison of all the sentences from the two articles. We do not take into account too short headlines (less than 3 words long) as they are unlikely to add to the representativeness of the resulting corpus.

The overall scheme is as follows: we iterate over all possible pairs of headlines (with the same date of publication) from different media agencies and calculate a similarity metric for each pair. Then the pairs with scores below the threshold value are pruned with the exception of a small portion of the necessary negative instances. The resulting dataset is evaluated by the annotators. Having the annotated data, we further adopt a supervised approach and optimize the unsupervised similarity metric.

3.2   Sentence Similarity Metric

To extract paraphrases, we use an unsupervised lexical similarity metric based on the one proposed by Fernando and Stevenson [15]:

$$sim(\vec{a}, \vec{b}) = \frac{\vec{a}W\vec{b}}{|\vec{a}||\vec{b}|},$$ (1)

where W is a similarity matrix and $\vec{a}$ and $\vec{b}$ – word vectors of the two sentences. Each element w_{ij} of the matrix represents the similarity between the words a_i and b_j. Diagonal elements obviously equal 1. Other elements equal 0 for different or $0 < \omega < 1$ for similar words. To capture lexical similarity, the authors use several metrics mainly based on the "$is\text{-}a$" hierarchy from WordNet [15].

In our research we use a matrix metric with lexical similarity scores based on the synonymy relation. As far as the source of synonymy relations is concerned, there exists a famous Russian dictionary of synonyms by Abramov [30] created over a century ago. Despite its numerous merits, the dictionary is deplorably outdated. The lack of modern synonymy resources has spurred a number of attempts at creating databases of synonyms. Although most resources are designed for practical purposes (rewriting texts in the web and automatically producing unique content), they can also be useful for NLP in general, and for our task in particular. We use one of such collections which consists of about 6 thousand articles. In fact, each article (a word and the list of its synonyms) can be considered a synset. For every pair of words we may further calculate the number of times they occur together in the same synset. In terms of information retrieval, every synset is a document, and it is known that semantically close are more likely to appear in the same documents than in the different ones. To compute lexical similarity, we use such metrics as normalized pointwise Mutual Information (npmi) [4], Dice coefficient [12] and Jaccard index [18].

Unlike the original metric from [15] (which uses WordNet relations), ours is calculated according to the list of scoring rules designed to capture not only synonymy relations but also conjugate words[2]:

---

[2] The latter might be of no importance for English, but they are essential for detecting Russian sentential paraphrases.

- Identical words starting with capital letters -> 1.2 score (a slight bias towards the simultaneous occurrence of the same named entities in the sentences).
- Identical words -> 1.
- Synonyms -> Npmi, Dice or Jaccard coefficient multiplied by 0.8.
- One of the words is a substring of the other -> the score equal to the length of the smaller word divided by the length of the larger word and multiplied by 0.7.
- The words have common prefix (at least 3 characters) -> the score equal to the prefix length divided by the length of the lesser word and multiplied by 0.
- Otherwise -> 0.

The original metric varies from 0 to 1. With our modifications it no longer satisfies this condition but it does not affect paraphrase extraction process. The scores (1.2, 1, 0.8, etc.) are obtained from the preliminary experiments conducted on the small subset of the corpus. Based on the results of these experiments, we select Jaccard index as the synonymy coefficient in our metric.

Let us calculate the metric for the two sentences from our dataset:

1. КНДР аннулировала договор о ненападении с Южной Кореей. /DPRK annulled the non-aggression treaty with South Korea/
2. КНДР вышла из соглашений о ненападении с Южной Кореей. /DPRK withdrew from the non-aggression agreement with South Korea/

We lemmatize the sentences using TreeTagger [23] and cut off auxiliary words. After these manipulations we represent the sentences as binary vectors (Fig. 1):

Word	аннулировать	выйти	договор	КНДР	Корея	ненападение	соглашение	Южный
	/annul/	/withdraw/	/treaty/	/DPRK/	/Korea/	/non-aggression/	/agreement/	/South/
S1 =	(1,	0,	1,	1,	1,	1,	0,	1)
S2 =	(0,	1,	0,	1,	1,	1,	1,	1)

Fig. 1. Example of word vectors for the two sentences

One can see that there are 4 overlapping words, 3 of them starting with an uppercase letter, and 2 synonyms: "соглашение" (agreement) and "договор" (treaty). The word "соглашение" occurs in 5 synsets in the dictionary of synonyms, while "договор" – only once (and their appear in one synset together). Jaccard index equals $1/(1 + 5 - 1) = 0.2$ for the two given words. This score is multiplied by the pruning coefficient: $0.2 * 0.8 = 0.16$. The similarity matrix for the two sentences is shown in Fig. 2.

According to (1), the resulting similarity score equals 0.763.

We apply the described metric to the pairs of article headlines and prune the ones with the Jaccard index-based similarity score below the empirically defined threshold value of 0.5.

Thus, our approach is based on the existing similarity metric, but according to our goal – the construction of paraphrase corpus for IE and TS – we introduce a list of scoring

	аннулировать	выйти	договор	КНДР	Корея	ненападение	соглашение	Южный
аннулировать	1	0	0	0	0	0	0	0
выйти	0	1	0	0	0	0	0	0
договор	0	0	1	0	0	0	0.16	0
КНДР	0	0	0	1.2	0	0	0	0
Корея	0	0	0	0	1.2	0	0	0
ненападение	0	0	0	0	0	1	0	0
соглашение	0	0	0.16	0	0	0	1	0
Южный	0	0	0	0	0	0	0	1.2

Fig. 2. Similarity matrix example

rules which capture the linguistic phenomena (synonymy relations, conjugate words, matching names entities) required for our corpus.

4 Corpus Annotation and Analysis

4.1 Annotation

Potential paraphrases with similarity metric values above the threshold are evaluated by the annotators. To obtain negative instances, we also include a portion of random sentence pairs with metric value below 0.5 in the corpus (roughly speaking, every fourth sentence pair in the corpus has a score below 0.5).

At the moment there are 5424 pairs of sentences in the corpus, with 5181 annotated pairs. Out of these 5181 pairs of sentences, we select 4758 pairs and include them in the corpus (by pruning inconsistent or potentially unreliable results).

We developed an online interface: http://paraphraser.ru/scorer for crowdsourced annotation. A user is shown two sentences at a time, and he/she is to decide whether the sentences convey the same meaning (class "1"), similar meanings (class "0") or different meanings (class "−1"). The users are advised to user their own judgement and intuition and are not given any specific instructions.

We try to make the annotation process less tedious by introducing an entertainment element: the users are shown various facts (about different events like the invention of something or the birth of a famous scientist/artist, etc.) and pictures at random intervals and are encouraged to annotate further.

It is well known that crowdsourcing poses a challenge concerning the reliability of the obtained results. To prune unreliable results, we only consider sentence pairs annotated by at least 3 users. If a paraphrase pair is annotated by less than 4 users, and two of them provide opposite judgments ("−1" and "1"), such pair is cut off. In future we also plan to involve expert linguists in the annotation process.

To assign a class to each sentence pair ("−1" for non-paraphrases, "0" for LPs and "1" – for PPs), we compute the median of all the scores given to the pair by the annotators. It can obviously take one of the following values: $\{-1, -0.5, 0, 0.5, 1\}$. As we would like to have only 3 classes, in case of ties we adopt a pessimistic strategy and round the value down to the previous integer (−0.5 is reduced to −1, 0.5 – to 0).

4.2 Paraphrase Classes

As stated earlier, we distinguish non-paraphrases, loose paraphrases (LPs) and precise paraphrases (PPs). Their distribution in our corpus is presented in Table 1.

Table 1. Distribution of paraphrase classes in the corpus

Paraphrase class	Number of instances	Percentage of instances
Non-paraphrases	1599	33.6 %
Loose paraphrases	1969	41.4 %
Precise paraphrases	1190	25 %
Total	*4758*	*100 %*

Although one cannot say that the dataset is severely unbalanced, there is a slight bias towards loose paraphrases. To analyze the differences between precise and loose paraphrases in the corpus we follow the approach adopted in [13]: we randomly select 100 PPs and 100 LPs and manually annotate them with the linguistic features:

- different content (sentences differ in words or phrases which carry additional information and make the sentences semantically different);
- different time (different grammar tenses are used to described the same event);
- context knowledge (sentences differ in the words or phrases, and added words/ phrases have no counterparts in the other sentences (see "different content"), but nevertheless it is clear from the context of the phrases that the same notion/event is being referred to, and that there is no semantic difference);
- metaphor (a metaphor takes place in one of the sentences);
- metonymy (sentences differ in some named entities, and one of these entities is used metonymically);
- numeral (sentence pairs differ in the representation of the same numerals or the number is rounded in one of the sentences);
- phrasal synonymy (sentences differ in the synonymous multiword expressions);
- reordering (sentences consist of the same words in different order);
- word-level synonymy (sentence pairs differ in the synonymous words);
- syntactic synonymy (the same information is expressed in the sentences using different constituents or the same constituents with different grammatical characteristics, e.g., a verbal phrase in active and passive voice respectfully).

Each pair of sentences can be annotated with more than one linguistic feature (e.g., syntactic synonymy is often accompanied by reordering and word-level synonymy). For each of the features we calculate the portion of sentence pairs it occurs in (see Table 2). These portions are calculated for PPs and LPs separately.

It can be seen that metaphor, metonymy and different representations of numbers are rare events in both types of paraphrases in our sample. While most (76 %) LPs differ in the meaning they convey (see "different content"), PPs are richer in word-, phrase- and syntactic-level synonymy. In fact, such results are quite predictable, and it is just what we expect these two paraphrase classes to be like. However, the portion of different content (18 %) among PPs (this feature is undesirable for PPs in our corpus) is not what one could call neglectable. Indeed, deciding on the semantic equivalence is a challenging task even for linguist experts, setting aside mere native speakers. Thus, in future we plan to involve experts' annotation to reduce the portion of semantically different phrases among PPs.

Table 2. Linguistic characteristics of precise and loose paraphrases

Feature	PP	LP	Feature	PP	LP
Context knowledge	32 %	25 %	Numeral	9 %	4 %
Different content	18 %	76 %	Phrasal synonymy	16 %	6 %
Different time	6 %	15 %	Reordering	17 %	10 %
Metaphor	0 %	1 %	Word-level synonymy	18 %	7 %
Metonymy	3 %	2 %	Syntactic synonymy	33 %	17 %

5   Evaluation

We evaluate the unsupervised metric used in the corpus construction by comparing it with the annotation results (see Table 3).

Table 3. Unsupervised paraphrase extraction: results

	Score above threshold	Score below threshold	Total
Precise paraphrases	1179	11	1190
Loose paraphrases	1919	50	1969
Non-paraphrases	763	836	1599
Total	3861	897	4758

As we do not distinguish between PPs and LPs when collecting sentence pairs using Jaccard-based metric, we only evaluate the quality of the metric in the task of classifying sentence pairs into similar (PPs and LPs) and different ones. Thus, its precision equals 80.24 %. We believe that the evaluation via traditional recall and F1 measures would be unreliable in our case. We do not focus on the collection of a balanced dataset with "proper" negative instances at the moment (approximately every fourth candidate is randomly selected as a potential negative instance), and recall and F1-score would overestimate our metric.

Our unsupervised metric extends the one used by Fernando and Stevenson [15]. They evaluated it against MSRP and achieved 75.2 % precision. Thus, our result seems promising although one should bear in mind that it only reflects the quality of classifying sentence pairs into 2 classes, when PPs and LPs are merged into one class.

Having obtained the annotated data, we optimized our metric: the threshold and the scores changed, Dice coefficient was selected instead of Jaccard index, and the overall performance improved, but due to space limitations we cannot give full details of the experiments here. At the moment we are working on a supervised approach towards paraphrase identification and train a classifier to distinguish between PPs and LPs, with the optimized similarity metric being used as one of the features. In future we intend to develop an approach which focuses on covering various paraphrase classes and linguistic phenomena [16, 21] because Russian is rich in these phenomena.

6   Conclusion: Future Work

In this paper we presented our work on the creation of a Russian sentential paraphrase corpus. The corpus consists of news headlines automatically collected from the Web and filtered using the unsupervised similarity metric. Such resource can be used in information extraction and text summarization. It can also serve as a training dataset for paraphrase identification models for Russian.

There are 4758 sentence pairs in the corpus at the moment, and it is freely available at our website: paraphraser.ru. All the pairs are being annotated using crowdsourcing via the website, and one of the three classes (non-paraphrase, loose paraphrase or precise paraphrase) is assigned to each pair of sentences. To obtain reliable data, we ensure that each pair of sentences in the corpus is annotated by at least 3 users and cut off inconsistent annotation. Evaluated against crowdsourced annotation, the similarity metric achieves 80.24 % precision at classifying paraphrases. Thus, it confirms our hypothesis that paraphrases can be extracted from the Russian news texts using methods based on lexical similarity.

Our further step aims at the development of paraphrase identification model and we are already working on it. This step includes using a better synonymy resource: Yet Another RussNet [5] (it is 8 times larger than our original one), and a dictionary of word formation families. We already use the optimized similarity metric in the paraphrase classifier and experiment with features based on semantic distributional models; other features are derived from the morphological characteristics, syntactic and semantic structure of the sentences. Thus, we intend to develop a fine-grained approach towards identifying paraphrases. As it might demand experts' annotation, it is one of our future work directions, along with the comparison of experts' annotation and the results of the automatic extraction of linguistic features. We acknowledge Saint-Petersburg State University for the research grant 30.38.305.2014.

References

1. Agirre, E., Cer, D., Diab, M., Gonzalez-Agirre, A., Guo W.: SEM 2013 shared task: semantic textual similarity. In: The Second Joint Conference on Lexical and Computational Semantics (2013)
2. Bannard, C., Callison-Burch, C.: Paraphrasing with bilingual parallel corpora. In: Proceedings of the 43rd Annual Meeting of the ACL, pp. 597–604 (2005)
3. Bernhard, D., Gurevych, I.: Answering learners' questions by retrieving question paraphrases from social Q&A sites. In: Proceedings of the ACL 2008 3rd Workshop on Innovative Use of NLP for Building Educational Applications, pp. 44–52 (2008)
4. Bouma, G.: Normalized (pointwise) mutual information in collocation extraction. In: Proceedings of the Biennial GSCL Conference (2009)

5. Braslavski, P., Ustalov, D., Mukhin, M.: A spinning wheel for YARN: user interface for a crowdsourced thesaurus. In: Proceedings of the Demonstrations at the 14th Conference of the European Chapter of the Association for Computational Linguistics, Gothenburg, Sweden, pp. 101–104 (2014)
6. Burrows, S., Potthast, M., Stein, B.: Paraphrase acquisition via crowdsourcing and machine learning. ACM Trans. Intell. Syst. Technol. **4**(3), 43 (2013)
7. Callison-Burch, C.: Paraphrasing and Translation. Institute for Communicating and Collaborative Systems, School of Informatics, University of Edinburgh (2007)
8. Chen, D.L., Dolan, W.B.: Collecting highly parallel data for paraphrase evaluation. In: Proceedings of the 49th Annual Meeting of the Association for Computational Linguistics, Portland, Oregon, USA, pp. 190–200 (2011)
9. Dzikovska, M.O., Nielsen, R., Brew, C., Leacock, C., Giampiccolo, D., Bentivogli, L., Clark, P., Dagan, I., Dang, H.T.: SemEval – 2013 Task 7: the joint student response analysis and 8th recognizing textual entailment challenge. In: Proceedings of the 7th International Workshop on Semantic Evaluation (SemEval 2013), Atlanta, Georgia, USA (2013)
10. Clough, P., Gaizauskas, R., Piao, S., Wilks, Y.: METER: MEasuring TExt Reuse. In: Isabelle, P. (ed.) Proceedings of the Fortieth Annual Meeting on Association for Computational Linguistics, Philadelphia, Pennsylvania, pp. 152–159 (2002)
11. Cohn, T., Callison-Burch, C., Lapata, M.: Constructing corpora for the development and evaluation of paraphrase systems. Comput. Linguist. Arch. **34**(4), 597–614 (2008)
12. Dice, L.R.: Measures of the amount of ecologic association between species. Ecology **26**(3), 297–302 (1945)
13. Dolan, W.B., Quirk, C., Brockett, C.: Unsupervised construction of large paraphrase corpora: exploiting massively parallel news sources. In: Proceedings of the 20th International Conference on Computational Linguistics, Geneva, Switzerland (2004)
14. Duboue, P.A., Chu-Carroll, J.: Answering the question you wish they had asked: the impact of paraphrasing for question answering. In: Proceedings of the Human Language Technology Conference of the North American Chapter of the ACL, New York, pp. 33–36 (2006)
15. Fernando, S., Stevenson, M.: A semantic similarity approach to paraphrase detection. In: Computational Linguistics UK (CLUK 2008) 11th Annual Research Colloqium (2008)
16. Fujita, A., Inui, K.: A class-oriented approach to building a paraphrase corpus. In: Proceedings of the Third International Workshop on Paraphrasing (2005)
17. Ganitkevitch, J., Callison-Burch, C.: The multilingual paraphrase database. In: Proceedings of the Ninth International Conference on Language Resources and Evaluation (LREC 2014). European Language Resources Association (ELRA), Reykjavik (2014)
18. Jaccard, P.: Étude Comparative de la Distribution Florale dans une Portion des Alpes et des Jura. Bulletin de la Société Vaudoise des Sciences Naturelles **37**, 547–579 (1901)
19. Knight, K., Marcu, D.: Summarization beyond sentence extraction: a probabilistic approach to sentence compression. Artif. Intell. **139**(1), 91–107 (2002)
20. McCarthy, Ph.M., McNamara, D.S.: The user-language paraphrase corpus. In: Cross-Disciplinary Advances in Applied Natural Language Processing: Issues and Approaches, pp. 73–89 (2008)
21. Rus, V., Banjade, R., Lintean, M.: On paraphrase identification corpora. In: Proceedings of the Ninth International Conference on Language Resources and Evaluation (LREC 2014), pp. 2422–2429. European Language Resources Association (ELRA), Reykjavik (2014)
22. Sanchez-Perez, M., Sidorov, G., Gelbukh, A.: The winning approach to text alignment for text reuse detection at PAN 2014. In: Cappellato, L., Ferro, N., Halvey, M., Kraaij, W. (eds.) Notebook for PAN at CLEF 2014. CEUR Workshop Proceedings, vol. 1180, pp. 1004–1011. CEUR-WS.org (2014). ISSN: 1613-0073

23. Schmid, H.: Improvements in part-of-speech tagging with an application to German. In: Proceedings of the ACL SIGDAT-Workshop, Dublin, Ireland (1995)

24. Shimohata, M., Sumita, E., Matsumoto, Y.: Building a paraphrase corpus for speech translation. In: Proceedings of the Fourth International Conference on Language Resources and Evaluation (LREC 2004). European Language Resources Association (ELRA), Lisbon (2004)

25. Shinyama, Y., Sekine, S.: Paraphrase acquisition for information extraction. In: Proceedings of the Second International Workshop on Paraphrasing, vol. 16, pp. 65–71 (2003)

26. Vila, M., Rodriguez, H., Marti, M.A.: WRPA: a system for relational paraphrase acquisition from wikipedia. Procesamiento del Lenguaje Nat. **45**, 11–19 (2010)

27. Wubben, S., van den Bosch, A., Krahmer, E., Marsi, E.: Clustering and matching headlines for automatic paraphrase acquisition. In: Proceedings of the 12th European Workshop on Natural Language Generation, Athens, Greece, pp. 122–125 (2009)

28. Xu, W., Ritter, A., Grishman, R.: Gathering and generating paraphrases from twitter with application to normalization. In: Proceedings of the Sixth Workshop on Building and Using Comparable Corpora, Sofia, Bulgaria, pp. 121–128 (2013)

29. Zhao, Sh., Lan, X., Liu, T., Li, Sh.: Application-driven statistical paraphrase generation. In: Proceedings of the 47th Annual Meeting of the ACL and the 4th IJCNLP of the AFNLP, Suntec, Singapore, pp. 834–842 (2009)

30. Abramov, N.: Slovar' russkih synonymov I shodnyh po smislu virazheniy, 7th edn. Russkie slovari, Moscow (1999)

Using Levenshtein Distance for Typical User Actions and Search Engine Switching Detection

Alexey Raskin$^{(\boxtimes)}$ and Petr Rudakov

National Research Nuclear University MEPhI,
Kashirskoe sh. 31, 115409 Moscow, Russia
a.a.raskin@gmail.com, rudakov-p@yandex.ru
http://www.mephi.ru/en

Abstract. This paper presents a new approach in automatic grouping of user search sessions. K-medoids clustering algorithm and Levenshtein distance function were used to group search sessions. We show that the groups obtained are meaningful and can be used to estimate the probability of user switching to another search engine. The proposed method was tested on real data provided by Yandex for 2012 Yandex Switching Detection Challenge and allowed for high AUC value (0.82 on internal tests). One more advantage of the presented approach is the possibility to visualize typical sequences of user action for simplified analyses of the data set.

Keywords: Levenshtein distance · Clustering · Sessions · Switching detection

1 Introduction

The problem of the allocation of typical user actions was raised in a fairly large number of papers [1,12,14]. In this article we propose a method of automatic partitioning of user sessions into groups. The method we suggest is straightforward and provides results that are easily interpreted; furthermore, it is also capable of determining levels of user satisfaction with the search results with relatively high fidelity.

Our work was inspired by [12] where Levenshtein distance was used for user sessions classification. We apply modified Levenshtein distance to compare sessions so that we could split the data into groups characterized by different probability of switching to another search engine. We assume that such clustering of a training set will help us determine the probability of the search engine switching during test sessions. Unlike the majority of approaches that consist of web sessions vectorization and subsequent application of conventional proximity measures to them [3,14], we choose to consider sessions as a whole. This approach allows us not to lose important information stored in the data structure and to have easily visualizable data.

© Springer International Publishing Switzerland 2016
P. Braslavski et al. (Eds.): RuSSIR 2015, CCIS 573, pp. 158–168, 2016.
DOI: 10.1007/978-3-319-41718-9_9

A practical application for partitioning browsing sessions may be found in estimating user satisfaction levels in general and detecting user switching between search engines in particular.

Users of searching engines (Google, Bing, Yandex, etc.) often use more than one searching engine. They can switch from one search engine to another during one search session. The problem of switching detection is important because it shows that the user is unsatisfied with search results and needs another search engine to find the information he/she needs. Only some switching steps can be easily detected (e.g. via information from the toolbar in the browser). When switching cannot be detected by standard ways, we need a detection technique based on characteristics of both, the user and the session.

For testing our algorithm we use data presented at Yandex Switching Detection Challenge 2012. The goal of the challenge was to predict for each of the test sessions how likely the user would be to switch to another search engine during the session.

2   Data Description

The data set includes 8,595,731 sessions, 10,139,547 unique queries, 49,029,185 unique URLs and 956,536 unique users. Two files are given in the dataset: a training file and a test one. The training set contains information about users that have done at least one switch in a period of 27 days. The test file consists of the sessions of the same users for the next three days with switching removed.

Each session is a sequence of queries, clicks or switching to another search engine with a number of characteristics for each step: time passed from session start, ID of the query and the list of URLs for queries ('Q'), information about time passed from session start and URL ID for click step by the user ('C'), as well as time passed from session start till the switching step ('S'). User ID and the day of the session are used to characterize the session itself.

We represent the data as a sequence with C-step (clicking step), S-step (switching step) and Q-step (querying step). Each element listed above uses the start of the session as its weight. Another set of data we use is the average number of switches per one session by every user. We do not use other information from datasets (Fig. 1).

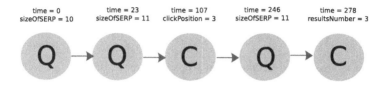

Fig. 1. Example of a session, where "sizeOfSERP" is the number of results on the search engine result page, "time" is the time marking the beginning of the step, and "clickPosition" is the numerical number of the link the user clicked.

3 Distance Function and Search Sessions

Levenshtein distance was chosen as a base for session comparison. As it was shown in [10], Levenshtein distance can be used to compare partially ordered sets. The results of the algorithm application were compared to other distance function algorithms [5, 15]. Below we provide the main principles used in comparing sessions that present more challenging cases of weighted sequence function application.

Suppose $R = (R_1, R_2, ..., R_I)$ and $R = (R'_1, R'_2, ..., R'_J)$ are sessions we need to compare, with $I \neq J$. The comparison should satisfy the following requirement: each element from the first sequence should be referenced with at least one element in the second sequence and vice versa, each element from the second sequence should be matched with at least one element from the first sequence. The correspondence between elements is not necessarily bijective (Fig. 2), it is in part due to $I \neq J$. The distance between sessions R and R' is a minimal sum of weights of all edges that connect vertices from different sessions.

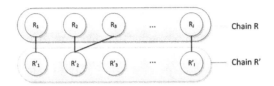

Fig. 2. Correspondence of vertices of two graphs

If we have two sequences of steps R and R' and normalized data for the times of those steps (T and T' respectively), then the distance between R and R' can be calculated using Eq. (1). This is a Levenshtein distance function with modified insertion, deletion and substitution operations. We used the well-known Wagner-Fischer [13] algorithm to calculate the distance function.

$$
L_{R,R'}(i,j) = \begin{cases} max(i,j) & if\, min(i,j) = 0 \\ min \begin{cases} L_{R,R'}(i, j-1) + 1 \\ L_{R,R'}(i-1, j) + 1 \\ L_{R,R'}(i-1, j-1) + \Delta(R(i), R'(j)) \end{cases} & ,else \end{cases}
$$

(1)

where Δ is a distance between two steps of sequences:

$$
\Delta(R(i), R'(j)) = \begin{cases} 1 + |T(i) - T'(j)| \, , \text{ if } R(i) \neq R'(j) \\ |T(i) - T'(j)| \quad\;\, , \text{ if } R(i) = R'(j) \end{cases}
$$

(2)

4   Typical Sequences Detection

4.1   Clustering Algorithm for Typical Sequences Detection

We use k-medoids algorithm with distance function described above to identify groups of similar sessions. This algorithm is used because of its low calculating requirements and easy parallel computing. Another reason for choosing K-medoids algorithm (instead of, for example, k-means) was to avoid calculating the mean value that cannot be determined. We find the problem of calculating the mean to be an area of future work and believe that the quality of clustering can be greatly improved by a more detailed study of this matter.

K-medoids algorithm is partitional and attempts to minimize the distance between points labeled to be in a cluster and a point designated as the center of that cluster. K-medoids algorithm chooses an existing data point as the center of a cluster (medoids or exemplars). The principle scheme of algortihm presented below:

1. Initiate start centers of clusters (medoids) μ_y for all clusters $y \in Y$
2. While y_i changes significantly:
   (a) $y_i \leftarrow \arg\max_{y \in Y} \rho(x_i, \mu_y), i = 1, ..., l$; (making clusters around medoids)
   (b) $\mu_{yj} \leftarrow \frac{\sum_{i=1}^{l}[y_i=y]f_j(x_i)}{\sum_{i=1}^{l}[y_i=y]}$ (updating medoids)

4.2   Typical Sessions Examples

We believe that the number of groups in the clustering (k-medoid algorithm requires a number of clusters as input parameter) must be defined based on specific goals established in the framework of user behavior analysis. To illustrate our point of view we provide the results of clustering our data in 14 groups, which has been accepted as the optimal solution to the problem of user switching to other search engines. Our reasons for such a choice will be further explained in Sect. 5.4. Some of the cases are listed and described below (each figure represents sessions of one user, each session, represented as a graph, has its own color assigned to it).

The simplest type of a user session is one query with some clicks and no specifying queries (Fig. 3). Such sessions are relatively short. One more example of a typical session is presented in Fig. 4 and shows a user who often specifies queries and switches between search engines if not satisfied. The most complicated case is a user with mixed different patterns of searching behavior (see Fig. 5).

5   Search Engine Switching Detection

5.1   Algorithm of Search Engine Switching Detection

Training set analyzed was divided into three parts:

1. Training set used to creating clusters

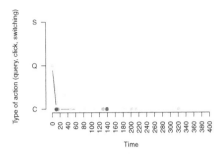

Fig. 3. Simple sessions with no specifying queries (Color figure online)

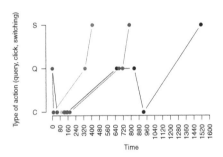

Fig. 4. Sessions with specifying queries and switching to another search engine (Color figure online)

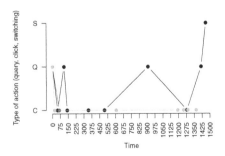

Fig. 5. Different search patterns mixed (Color figure online)

2. Cross-validation set used to evaluate the optimal number of clusters and to avoid overfitting
3. Test set for internal AUC calculating

The whole algorithm of detecting sessions with high probability of switching to another search engine is presented below. We repeat these steps increasing the number of clusters until the quality of the prediction model (AUC value) begins to decrease (or stops to increase):

1. Clustering on training set with k-medoids algorithm and presented distance function.
2. Calculating the frequency of switching to another search engine for each cluster we get
3. Assessing the quality of clustering and the capability of the developed model to predict user switching to another search engine based on the validation sample:
   (a) The nearest cluster is determined for each of the sessions from the cross-validation dataset.
   (b) The frequency of switching to another search engine among the sessions from the nearest cluster was used to assess the probability of switching to another search engine in the session
   (c) The probability was calculated as a product of the frequency of switching in the nearest cluster and the frequency of switching to a different search engine for all sessions of a user.
   (d) Prediction model AUC properties were calculated based on available and valid cases of user switching to another search engine (the switch that really happened).

5.2   Results

AUC (Area Under Curve) measure [7] was used to evaluate the results and efficiency of our algorithm. This measure is used for evaluation of predictors and represents "the probability that a classifier will rank a randomly chosen positive instance higher than a randomly chosen negative instance" [2].

If the results of the classifier are given as a ranked list based on the beliefs that each instance is positive, we can use the next formula (according to [7]) to calculate AUC:

$$AUC = \frac{(S_0 - n_0(n_0 + 1)/2)}{n_0 n_1}$$

where $S_0 = \sum(r_i)$, r_i are the ranks of truly positive examples in the results list, and n_0, n_1 are the number of positive and negative examples respectively. AUC was shown to be better (more discriminating) measure than accuracy.

AUC values on internal tests and on Yandex testing system provided for the challenge turned out to be different. It became apparent that the data from the test period are slightly different from the data for the training period.

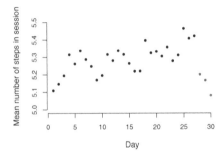

Fig. 6. Average length of sessions (in steps) in training period (black) and test period (red) (Color figure online)

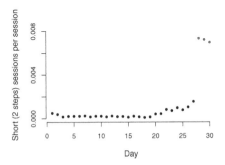

Fig. 7. Number of short sessions per session (for users who always switched to another search engine) (Color figure online)

For example, on average, sessions from test period are shorter than sessions for the users from training period (Fig. 6). One more difference between the test and the training sets lies in the behavior of the users who (according to statistics of the test period) always switched to a different search engine. We compared sequences of actions of such users in the test set and the training set and found that in the test period users had significantly more two-step sessions (Fig. 7). We cannot know the reason for such a difference and can only suggest it is due to the timing of the end of training and test periods, they both fell on a holiday period.

The average value of AUC we received is 0.775 (according to Yandex test system) and 0.82 (according to internal tests). The last value is in the top ten results in the Challenge, and this is the only approach (among published [4, 8, 11]) that does not use vectorization and preserves the structure of data for future analyses and visualization. The approach described in [4] produces the AUC value of 0.844. Its main idea is to allocate 44 features and use neural network on the vectorized data set. It means converting data into a 44-dimensional space which is difficult for both: analysis and visualization. Authors of [11] suggest applying pGBRT implementation of gradient tree boosting and 414 features of the data set. It should be noted that these features include n-grams that reflect the most common subsequence of user actions, even though they do not reflect the whole sequence of actions. This approach gets the AUC value of 0.849. In [8] there were used a number of techniques like online Bayesian probit regression and support vector regression. They also use n-gram technique, but in spite of very high result (AUC is equal to 0.843) the authors are going to "conduct other sequence learning models (e.g., Hidden Markov models) to explore the sequence properties from the dataset" in the future.

5.3 Stability of Clustering

Despite the fact that clustering process includes random definition of centroids, the results are quite stable. To reduce the risk of bad clustering results we initialized centroids by picking points (sessions) that are as far away from one another as possible [9]. We have to use heuristic function for estimating distance between objects to reduce computational complexity. The difference between lengths of sessions is accepted as a heuristic function.

The experiment was repeated and AUC value was measured 20 times (on one set of data and one number of clusters). According to Shapiro-Wilk normality test we can consider AUC values as normally distributed data ($W = 0.942, pvalue = 0.3431$), with 0.775 as the mean value (according to Yandex test system) and 0.002 as its standard deviation. Our choice fell on Shapiro-Wilks test because of the small number of observations.

Our findings established that even a quarter of the training set is enough to obtain a stable result in switching prediction, which means that even five days of data is enough for form groups of typical sessions and predicting switching to another search engine.

5.4 Dependence on Number of Clusters

We have defined the range of clusters with the highest AUC measure. It is a rather large range of 14–30 clusters (Fig. 8). In our opinion, there is a reason for such a large range: sessions with similar structure (and similar probability of switching to another search engine) are merged into one cluster when splitting into 14 clusters. With the increase in the number of clusters, there are more clusters for the data to be split into. However, the latter does not affect the predictive power of the user switching estimation because of the similarity of the session structure and similar probabilities of switching. AUC value decreases when the amount of clusters exceeds 30. The difference between AUC values for the number of clusters in the range defined above is very small, consequently, we chose 14 clusters as the minimum number of clusters with a high quality of prediction.

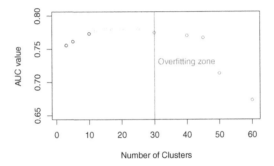

Fig. 8. Quality of prediction (AUC value) given number of clusters

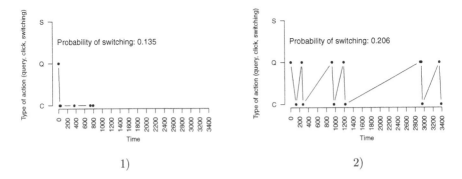

Fig. 9. Medoids of some defined clusters used to predict switching to another search engine (with low probability of switching)

5.5   Ability of Typical Session Visualization

Our approach also has an additional advantage of establishing a typical sequence of actions (a typical user behavior). This typical sequence of user actions can be represented through centroids of obtained clusters.

Our cluster analyses identified two groups of users: one group was character- ized by short sessions and subsequent Q-steps and C-steps (Fig. 9(1)); the other group of users had a lot query entries and clicks which followed one after another (Fig. 9(2)). During such search sessions the probability of switching is quite low because users either have low interest in search results, or their queries are very simple and they can find the relevant link easily.

Figure 10(1) shows how the probability of switching increases if numerous queries are made to specify a previous query. In certain cases a user tries to specify his query with no clicking steps (possibly because the user sees that the query produces irrelevant links) (Fig. 10(1)). In the remaining cases our user refines the query after the initial search results are analyzed (Fig. 10(2)).

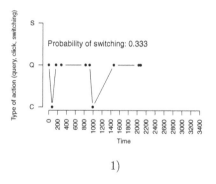
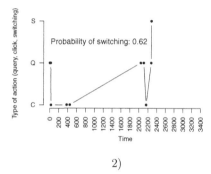

1)   2)

Fig. 10. Medoids of some defined clusters used to predict switching to another search engine (with average or high probability of switching)

6   Conclusion

Modified Levenshtein algorithm can be used as distance function for clustering of search engine user sessions. Obtained clusters can be in turn used to predict switching to another search engine. Our proposed algorithm was tested on data provided by Yandex for 2012 Yandex Switching Detection Challenge and achieved high results: the AUC value is 0.775 (according to Yandex test system) and 0.82 (according to internal tests). The second value is in the top ten results of the Challenge.

In our opinion the main benefit of this approach is the preservation of data structure. Unlike other approaches, the proposed algorithm creates clusters and does not vectorize clusters and centroids, hence they can be easily processed by analysts and in some cases easily visualized. The main disadvantage of the approach is the inability to identify the mean value for all values in the cluster. In this study we have resolved the problem by using k-medoids algorithm to avoid calculating the mean.

In the future we would like to investigate the possibility of calculating the mean of sessions to increase predictive quality and define typical sequences of user steps.

References

1. Ageev, M., Guo, Q., Lagun, D., Agichtein, E.: Find it if you can: a game for modeling different types of web search success using interaction data. In: Proceedings of the 34th International ACM SIGIR Conference on Research and Development in Information Retrieval (SIGIR 2011), pp. 345–354. ACM, New York (2011)
2. Fawcett, T.: An introduction to ROC analysis. Pattern Recogn. Lett. **27**(8), 861–874 (2006)
3. Heath, A.P., White, R.W.: Defection detection: predicting search engine switching. In: Proceedings of the 17th International Conference on World Wide Web, pp. 1173–1174. ACM, New York (2008)

4. Kalinin, P.: Neural networks applied to switching prediction. Voronezh State University. In: The Proceedings of the 6th ACM International Conference on Web Search and Data Mining (WSDM 2013), Rome, Italy, February 2013

5. Kendall, M., Gibbons, J.D.: Rank Correlation Methods, 5th edn. Oxford University Press, Oxford (1990)

6. Levenshtein, V.: Binary codes capable of correcting deletions, insertions and reversals. Sov. Phys. Dokl. **10**, 707 (1966)

7. Ling, C., Huang, J., Zhang, H.: AUC: a statistically consistent and more discriminating measure than accuracy. In: International Joint Conference on Artificial Intelligence, vol. 18, pp. 519–526 (2003)

8. Yan, Q., Wang, X., Qiang, X., Kong, D., Bickson, D., Yuan, Q., Yang, Q.: Predicting search engine switching in WSCD 2013 challenge. In: Workshop on Web Search Click Data (WSCD), Rome, Italy, February 2013

9. Rajaraman, A.: Jeffrey David Ullman: Mining of Massive Datasets. Cambridge University Press, New York (2011)

10. Raskin, A.: Comparison of partial orders clustering techniques. Proc. ISP RAS **26**(4), 91–98 (2014)

11. Savenkov, D., Dmitry, L., Liu, Q.: Search engine switching detection based on user personal preferences and behavior patterns. In: Proceedings of the 36th International ACM SIGIR Conference on Research and Development in Information Retrieval, Dublin, Ireland, 28 July–01 August 2013 (2013)

12. Scherbina, A., Kuznetsov, S.: Clustering of web sessions using levenshtein metric. In: Perner, P. (ed.) ICDM 2004. LNCS (LNAI), vol. 3275, pp. 127–133. Springer, Heidelberg (2004)

13. Wagner, R., Fischer, M.: The string-to-string correction problem. J. ACM. **21**(1), 168–173 (1974)

14. White, R.W., Dumais, S.T.: Characterizing and predicting search engine switching behavior. In: Proceedings of the 18th ACM Conference on Information and Knowledge Management (CIKM 2009), pp. 87–96. ACM, New York (2009)

15. Ukkonen, A.: Clustering algorithms for chains. J. Mach. Learn. Res. **12**, 1389–1423 (2011)

Detecting Opinion Polarisation on Twitter by Constructing Pseudo-Bimodal Networks of Mentions and Retweets

Igor Zakhlebin[1], Aleksandr Semenov[1],
Alexander Tolmach[2], and Sergey Nikolenko[3,4,5(✉)]

[1] International Laboratory for Applied Network Research,
National Research University Higher School of Economics,
Moscow, Russian Federation
{izakhlebin,avsemenov}@hse.ru
[2] Institute of Sociology, Russian Academy of Sciences, Moscow, Russia
quatsch.ad@gmail.com
[3] Kazan (Volga Region) Federal University, Kazan, Russia
sergey@logic.pdmi.ras.ru
[4] Laboratory for Internet Studies,
National Research University – Higher School of Economics,
St. Petersburg, Russia
[5] Steklov Institute of Mathematics at St. Petersburg, St. Petersburg, Russia

Abstract. We present a novel approach to analyze and visualize opinion polarisation on Twitter based on graph features of communication networks extracted from tweets. We show that opinion polarisation can be legibly observed on unimodal projections of artificially created bimodal networks, where the most popular users in retweet and mention networks are considered nodes of the second mode. For this purpose, we select a subset of top users based on their PageRank values and assign them to be the second mode in our networks, thus called pseudo-bimodal. After projecting them onto the set of "bottom" users and vice versa, we get unimodal networks with more distinct clusters and visually coherent community separation. We developed our approach on a dataset gathered during the Russian protest meetings on 24th of December, 2011 and tested it on another dataset by Conover [13] used to analyze political polarisation, showing that our approach not only works well on our data but also improves the results from previous research on that phenomena.

Keywords: Twitter · Opinion polarisation · Two-mode networks · Community detection

1 Introduction

Twitter has become one of the most popular social networking services among researchers due to the open nature of its communication and relatively easy access to its data via the API (Application Programming Interface). The scope

© Springer International Publishing Switzerland 2016
P. Braslavski et al. (Eds.): RuSSIR 2015, CCIS 573, pp. 169–178, 2016.
DOI: 10.1007/978-3-319-41718-9_10

of previous Twitter-related research includes detection of the users' psychological features [14], spread of diseases [2] and response to natural disasters [24], analysis of financial markets [6], electoral predictions [15], and marketing campaigns [10]. One of the most scrutinised directions of study, however, concerns the protest movements in Twitter like the "#occupy" movement [4] or the so-called Twitter revolutions in the Middle East [12].

In this direction, researchers compared language use in Egypt and Libya [7], analysed types of actors [18] and measured the recruitment patterns and dynamics [19]. All these topics are related to a general question about political polarisation on Twitter because it can be used by all sides of the conflict in question to promote their point of view, strengthen their group identity and discriminate the opposite sides [25].

One of the most famous examples of political polarisation on the Internet was presented in [1]. The authors used a network approach to analyze hyperlink patterns among US political blogs during the presidential campaign of 2004 and demonstrated highly separated nature of pro-Republican and pro-Democrat parts of the blogosphere. Although there is other evidence that hyperlinks in blogs can serve as a signal of ideological affiliation [20], applying this approach to Twitter might not work well because hyperlinks are used sporadically, don't stay visible for long as the timeline fills with another updates, and can be used by all sides of the conflict in both positive and negative way. This suggests that analysis and visualisation of networks based on hyperlinks will not result in a clear picture of community structure. Some of these and other important differences between hyperlink usage in blogs and Twitter have been discussed in [9].

Another approach to detecting the stance of Twitter users on an issue of interest is based on the usage of keywords or hashtags related to that issue. In practice, searching for a hashtag is one of the most popular ways to get a sample of tweets [8]; however, it might introduce its own problems with the bias of that sample. For example, data gathered from trending hashtags during other protest meetings in Russia showed that these hashtags form two distinct clusters with pro-opposition and anti-opposition tweets [22]. However, both hashtags and clusters they represent contain words with clearly negative connotations and do not show more cautious or casual opinions on the events that use more neutral synonyms like "meeting", "march" etc.

We have not been able to find any kind of network analysis on the resulting dataset in previous work, although network analysis had proven to be useful in similar situations. For example, in [13] the authors gathered hashtags associated with Democratic and Republican parties for several weeks during the midterm elections to the US Congress in 2010. It goes without saying that political polarisation discovered in this paper was not a huge surprise; however, it was discovered only for the retweets while networks of mentions were more homogeneous with a low modularity score of 0.17.

Although these results look well-grounded and reasonable, there is one caveat in the general approach. Gathering trending or politically biased keywords may result in biased and polarised datasets. But when we need to cover the entire spectrum of opinions, we need to gather as much neutral keywords or hashtags as possible. In this work, we suggest an approach to this problem similar in spirit to the work [17]. In that paper, the authors classified ordinary users of Twitter and media outlets via the politicians whom these users follow on Twitter. The rationale behind this is simple – the number of prominent politicians and Congress members is limited and their position on the political spectrum is well-known. Therefore, it is reasonable to assume that their followers share that position and hence put the main media outlets on this continuum through their profiles.

Since Twitter had changed its API limits for gathering data on followers, making it almost impossible to build large graphs on that type of relationship, we decided to apply this logic to the networks built from retweets and mentions. Previous research demonstrated that users tend to retweet those whose ideas they share [9,13,25] and that there are very few popular, prominent, and central users [3] who can serve as such opinion leaders and whose influence does not depend on their followers count alone [11]. This allows us to assume that these "influencers" might be seen as a special type, or, in network terms, a "second mode" of users. Hence, we can analyze unimodal networks of user communication as if they were bimodal networks, artificially separating top (most popular) users into a second mode[1]. We analyze these networks as if they were "normal" bimodal networks by projecting sets of both "top" and "bottom" users on each other to obtain two separate unimodal networks for "top" and "bottom" users. The results show that this approach leads to more distinct clusters in the Twitter mention networks compared to standard analysis of bimodal networks constructed from hashtags and hyperlinks.

The rest of the paper is organised as follows. In the next section, we describe the network features of communication on Twitter with an emphasis on data acquisition and network extraction methods. In Sect. 3, we describe our approach to the polarisation discovery via unimodal projections of pseudo-bimodal networks. In Sect. 4, we describe our dataset and its background. Section 5 demonstrates the results of our analysis. Finally, in conclusion we discuss possible limitations of our results and how further work can help avoid them and improve our approach.

2   Communication Networks on Twitter

On Twitter, users communicate by posting short public text statuses, often containing hyperlinks and pictures, called "tweets". To indicate that a tweet deals with a certain topic, users insert "hashtags" in their tweets which consist of the number sign ("#") followed by an alphanumeric combination denoting the topic (like "#tcot").

There are three basic ways how users can interact. Users can "mention" particular persons by including the recipients user name prefixed with "@" sign in the tweet (like "@navalny"). Such a combination is called a Twitter "handle".

---

[1] The term pseudo-bimodal networks is based on the previously introduced notion of pseudo-tricluster in two paired bimodal networks [16].

To show support, users can quote tweets to their timelines, thus sharing them with their subscribers; this is called a "retweet." Retweets start with "RT" and a handle. A tweet immediately starting with a handle is called "reply" and is considered to be a direct message from one user to another.

Although there are substantial differences between mentions, retweets, and replies, we define a mention as any occurrence of a user's handle in a tweet. We did it mostly because other authors rarely analyze reply networks on their own and the most common type of networks used in the analysis are retweets and mentions. Moreover, the dataset from [13], which we use as a test case for our approach, follows that classification too. Thus, in this paper mentions formally include both replies and retweets.

As any mention denotes a reference from one user to another, two types of directed networks were constructed from it: mentions and retweets. Nodes in these networks stand for the users and edges denote the chosen type of interaction; they are directed from the users who posted tweets to the users being mentioned in them.

The fact that any user can see any tweet lets any user freely gain popularity on Twitter. As a consequence, some individuals can gain influence comparable to and even surpassing that of organisations like news companies represented on the service. Such important influencers generally include politicians, individual bloggers, and celebrities.

To measure user influence in terms of network topology, let us consider simple measures of their centrality. A certain set of nodes has small out-degrees and large in-degrees. Those users produce mostly original tweets that get mentioned often; in what follows we call them "top users". For most other nodes, the out-degrees are larger than in-degrees, which is commonly interpreted as their activity score. These are the users that retweet others but do not get retweeted often; in what follows we call them "ordinary users". Another network metric, which demonstrates the intuition that prominence of a user is defined by centrality of his peers, is PageRank [23] and it is used here in further analysis.

Another observation on our data is that top users rarely mention each other. Most interactions happen between the ordinary users and the top ones. Thus, if we try to pull the network of mentions between top users, it will be too sparse to search for communities in it. On the other hand, the full network of interactions may be too strongly interconnected to effectively partition it as well [13].

These observations allow us to cluster ordinary users according to which top users they retweet and mention. To reiterate, our approach is based on the following assumptions:

- there exists a small group of users in the network with high PageRanks;
- they rarely interact with each other and with ordinary users;
- ordinary users tend to mention mostly top users with whose opinions they agree;
- users from both sets predominantly belong to one group each according to their opinions.

3   Method for Detecting Opinion Polarisation

Our proposed algorithm receives a directed communication network $G = (V, E)$ as an input, where V and E are, respectively, its sets of nodes and edges. The algorithm consists of the following operations sequentially performed on G:

1. *Select a set of top users for some threshold k.*

To separate top users from the ordinary ones, we sort the corresponding nodes by their PageRank values and simply select k nodes with highest values. This splits them in two disjoint sets: top users V_T and ordinary users V_B, $V = V_T \cup V_B$, $V_T \cap V_B = \emptyset$.

2. *Make the network bimodal.*

To complete separation of top users into the second mode, all edges in E between nodes of the same set $(V_T \times V_T \cup V_B \times V_B)$ are removed. This produces a bipartite graph which we call the *pseudo-bimodal* network, $G^* = (V_T, V_B, E^*), E^* \subset E$. Its edges show how did the ordinary users mention top ones and vice versa. It is subsequently analysed as if it were a regular bimodal network.

3. *Project the pseudo-bimodal network onto one of its node sets.*

Having constructed the pseudo-bimodal network G^*, we can either study ordinary users by their mutual connections to the top ones or study top users by intersecting their audiences, i.e., subsets of ordinary users mentioning them. For that purpose, we use Newman's two-mode projection method [21] to get a unimodal undirected weighted network built on a selected set of users (that is, a projection of the network on the set V_T or V_B). We begin by defining this process for the projection of G^* on the set of top users V_T. For a pair of nodes $i, j \in V_T$, $L_{i,j}$ denotes the set of nodes connected to both i and j, $L_{i,j} = \{l \in V_B | (l, i) \in E^*, (l, j) \in E^*\}$. Both i and j will occur in the projected one-mode network, and they will be connected iff the set $L_{i,j}$ is nonempty. The edge between i and j is weighted as $w_{i,j} = \sum_{l \in L_{i,j}} \frac{1}{k_l - 1}$, where $k_l = |\{i \in V_T | (l, i) \in E^*\}|$. Projection on the set V_B is done similarly.

4. *Perform community detection on the resulting one-mode network.*

The one-mode network obtained on the previous step is expected to have a more expressed structure with more tight-linked communities and higher modularity. To partition the users into groups with presumably similar political biases we use the Louvain graph clustering method [5], one of the best known and widely used methods for community detection, on the resulting one-mode network. The Louvain method looks for a graph partitioning that maximizes *modularity*, i.e., density of links inside communities compared to links between communities. Modularity is defined as $Q = \frac{1}{2m} \sum_{i,j} (1 - \frac{k_i k_j}{2m}) \delta(c_i, c_j)$, where m is the total number of edges in the graph, k_i is a degree of node i, c_i and c_j are the communities of the nodes, and δ is a delta function ($\delta(c_i, c_j) = 1$ if $c_i = c_j$ and 0 otherwise). Q varies between 0 and 1, with 1 corresponding to a perfect separation of nodes, i.e., no edges between different clusters.

As a result, we find community structures among top and ordinary users. These community structures for ordinary users represent how often they retweet and/or mention the same top users (i.e., whether they follow the same issues);

Table 1. Descriptive statistics for the datasets

Name	Data sources	Number of			
		Users	Tweets	Mentions	Retweets
24th December, Russia	Streaming and Firehose API	3,485	24,378	12,725	6,529
U.S. Elections	Firehose API	45,000	250,000	77,920	61,157

for top users they show how often they are mentioned by the same ordinary users (i.e., how much their audiences overlap). This leads to a different and more pronounced community structure, as we will see in practical examples below.

4   Datasets

We have used the following datasets related to political polarisation.

1. *Meetings on December 24th, 2011 in Russia.*

As a main source of data we used tweets on pro-government and protest political events happened in Moscow during December 24th, 2011 on Poklonnaya Gora and Prospekt Sakharova. We have collected them using Twitter's Streaming API and Firehose. The first is Twitter's own free source of data, which contains a 1% sample of all the tweets. Firehose is a full stream of tweets with a cap of 500,000 tweets per hour, provided on a commercial basis by DataSift (www. datasift.com). To collect only tweets that refer to political events, we filtered them according to hashtag "24дек" ("#24dec", short of Russian "December 24th"), which was heavily used by both sides during that day and did not favor any particular position. Thus, we gathered 24,378 tweets from 3,485 unique users with 12,725 mentions, 6,529 of which were retweets.

2. *U.S. midterm elections to the Congress in 2010.*

This dataset was used in the work [13] and has been made public. We use it to test our approach on similar data from a similar context because it is one of the rare cases of publicly available datasets from Twitter. Descriptive statistics of both datasets are provided in Table 1.

5   Results

From these datasets, we have constructed:

- two standard bimodal networks: (a) network of users and hashtags which they used in their tweets (hashtag network), and (b) network of users and domains to which these users referred via hyperlinks (domain network);
- two unimodal networks with respect to retweets and mentions.

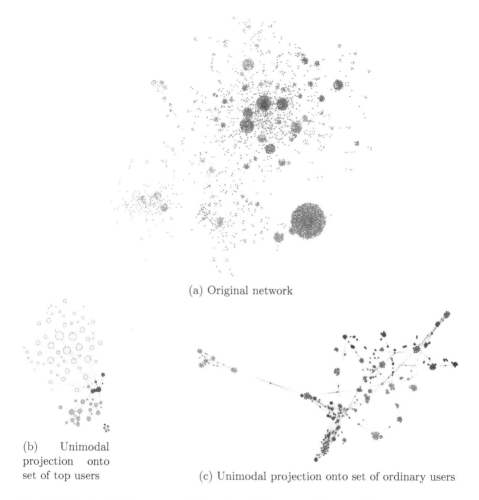

(a) Original network

(b)   Unimodal
projection   onto
set of top users

(c) Unimodal projection onto set of ordinary users

Fig. 1. Variants of retweet network based on 24 Dec. 2011 dataset (a) Original network (b) Unimodal projection onto set of top users (c) Unimodal projection onto set of ordinary users

We analysed the latter two as bimodal networks. To transform communication networks into bimodal ones, we chose the top 100 users by PageRank as the second mode in the network and projected the result onto a unimodal network with Newman's method [21] to see how they are connected among each other via ordinary users who reply to and/or retweet them. Then we clustered each network with the Louvain method and used its modularity coefficient, which is one of the most common measures of clustering quality on networks.

As expected, results from projections of bimodal networks constructed from hashtags returned the least readable clusters of users. With modularity score of 0.122, this network contained neutral hashtags with dates, names of cities, and other non-polarising keywords. The unimodal projection of the network constructed from URLs was a bit better in terms of modularity (0.485) but contained such hubs as youtube, livejournal, twitter, facebook, and http://vk.com, which once again were neutral in terms of possible content and usage. We view these results as evidence for the fact that unimodal networks from hashtags and URLs do not detect clusters particularly well in our case.

After the clustering, for our dataset collected during the events of 24 Dec. 2011, we get unimodal networks of retweets presented on Fig. 1 (for $k = 100$). It is clear that projected graphs are much better structured: connections are dense inside the clusters and sparse between them. What is even more surprising, this method also works for networks of mentions which are usually considered to be more homogeneous [9,13]: both in our dataset and the test data from [13] the clusters obtained after projection are much better defined. Figure 2 shows how modularity of both unimodal networks changes with the percentage of top users; observe that even for a small number of top users (left part of the graph), which, naturally, do not form a modular graph, the modularity of the "ordinary" part of the graph increases.

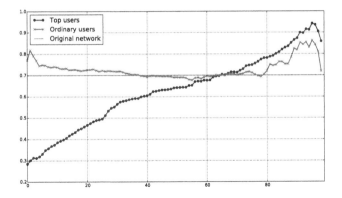

Fig. 2. Modularity as a function of the cutoff value for the Conover dataset.

6   Conclusion

We have proposed an approach to detect and explore opinion polarisation in Twitter communication networks, which leads to better defined clusters of users than methods employed in previous works. We have shown that our method works not only on our dataset, but also on a classical dataset previously used in literature, improving the quality of clustering.

However, our approach has some limitations. First, it would be good to have a mathematical proof that our results are not an artifact of bimodal networks and projection methods; this concern might also be solved via simulations or an analytic solution. Second, currently we have only analysed data with political origins, which are polarised by nature. Perhaps, in more homogeneous contexts such as tweets from scientific conferences or pop culture entertaining events this approach will not work so well. Hence, we need to test our approach on more datasets from different contexts on Twitter and maybe in other social media and domain areas. Third, although we have managed to improve upon the results of [13], we cannot verify all conclusions since not all information on the dataset has been provided by authors and also because of our lack of substantial knowledge of US political situation both offline and in Twitter.

Therefore, as further work we plan to test the approach analytically and via simulations, try different centrality measures (eigenvector, HITS), projection methods, and cut-off values, use semi-supervised annotation of tweet texts and labeling to attain a ground truth about existing opinions and clusters, add non-political and non-Twitter datasets into the analysis, and, finally, test how user preferences persist through time (across several datasets about Russian political events) and see if user separation remains stable.

Acknowledgements. This paper was prepared within the framework of a subsidy granted to HSE by the Government of Russian Federation for implementation of the Global Competitiveness Program. The work of Sergey Nikolenko was supported by the Russian Science Foundation grant no. 15-11-10019.

References

1. Adamic, L.A., Glance, N.: The political blogosphere and the 2004 U.S. election: divided they blog. In: Proceedings of the 3rd International Workshop on Link Discovery, LinkKDD 2005, pp. 36–43. ACM, New York, NY, USA (2005). http://doi.acm.org/10.1145/1134271.1134277
2. Aramaki, E., Maskawa, S., Morita, M.: Twitter catches the flu: detecting influenza epidemics using twitter. In: Proceedings of the Conference on Empirical Methods in Natural Language Processing, pp. 1568–1576. Association for Computational Linguistics (2011)
3. Bakshy, E., Hofman, J.M., Mason, W.A., Watts, D.J.: Everyone's an influencer. In: Proceedings of the Fourth ACM International Conference on Web Search and Data Mining - WSDM 2011, p. 65. ACM Press, New York, New York, USA (2011). http://portal.acm.org/citation.cfm?doid=1935826.1935845
4. Bennett, W.L., Johnson, C.N.: A model of crowd-enabled organization: theory and methods for understanding the role of Twitter in the occupy protests. Int. J. Commun. **8**, 646–672 (2014)
5. Blondel, V.D., Guillaume, J.L., Lambiotte, R., Lefebvre, E.: Fast unfolding of communities in large networks. J. Stat. Mech.: Theory Exp. **2008**(10), P10008 (2008)
6. Bollen, J., Mao, H., Zeng, X.J.: Twitter mood predicts the stock market (2010). arXiv preprint arxiv:1010.3003
7. Bruns, A., Highfield, T., Burgess, J.: The Arab spring and social media audiences: English and Arabic Twitter users and their networks. Am. Behav. Sci. **57**(7), 871–898 (2013). http://abs.sagepub.com/cgi/doi/10.1177/0002764213479374
8. Bruns, A., Burgess, J.E.: The use of Twitter hashtags in the formation of ad hoc publics. In: 6th European Consortium for Political Research General Conference (2011)

9. Bruns, A., Highfield, T.: Political networks on Twitter. Inf. Commun. Soc. **16**(5), 667–691 (2013). http://www.tandfonline.com/doi/abs/10.1080/1369118X.2013.782328

10. Bulearca, M., Bulearca, S.: Twitter: a viable marketing tool for SMEs. Glob. Bus. Manag. Res.: Int. J. **2**(4), 296–309 (2010)

11. Cha, M., Haddai, H., Benevenuto, F., Gummadi, K.P.: Measuring user influence in Twitter: the million follower fallacy. In: International AAAI Conference on Weblogs and Social Media (2010)

12. Comunello, F., Anzera, G.: Will the revolution be tweeted? A conceptual framework for understanding the social media and the Arab Spring. Islam Christ. Muslim Relat. **23**(4), 453–470 (2012). http://www.tandfonline.com/doi/abs/10.1080/09596410.2012.712435

13. Conover, M., Ratkiewicz, J., Francisco, M., Gonalves, B., Menczer, F., Flammini, A.: Political polarization on Twitter. In: ICWSM (2011). http://www.aaai.org/ocs/index.php/ICWSM/ICWSM11/paper/viewFile/2847

14. Coppersmith, G., Dredze, M., Harman, C.: Quantifying mental health signals in Twitter. In: Proceedings of ACL Workshop on Computational Linguistics and Clinical Psychology, Association for Computational Linguistics (2014)

15. Gayo-Avello, D.: A meta-analysis of state-of-the-art electoral prediction from Twitter data. Soc. Sci. Comput. Rev. **31**(6), 649–679 (2013). http://ssc.sagepub.com/cgi/doi/10.1177/0894439313493979

16. Gnatyshak, D., Ignatov, D.I., Semenov, A., Poelmans, J.: Gaining insight in social networks with biclustering and triclustering. In: Aseeva, N., Babkin, E., Kozyrev, O. (eds.) BIR 2012. LNBIP, vol. 128, pp. 162–171. Springer, Heidelberg (2012)

17. Golbeck, J., Hansen, D.: A method for computing political preference among Twitter followers. Soc. Netw. **36**, 177–184 (2014). http://linkinghub.elsevier.com/retrieve/pii/S0378873313000683

18. Gonzalez-Bailon, S., Borge-Holthoefer, J., Moreno, Y.: Broadcasters and hidden influentials in online protest diffusion. Am. Behav. Sci. **57**(7), 943–965 (2013). http://abs.sagepub.com/cgi/doi/10.1177/0002764213479371

19. González-Bailón, S., Borge-Holthoefer, J., Rivero, A., Moreno, Y.: The dynamics of protest recruitment through an online network. Sci. Rep. **1**, 197 (2011). http://www.pubmedcentral.nih.gov/articlerender.fcgi?artid=3240992&tool=pmcentrez&rendertype=abstract

20. Highfield, T.: Mapping intermedia news flows: topical discussions in the Australian and French political blogospheres. Ph.D. thesis, Queensland University of Technology (2011)

21. Newman, M.E.J.: Scientific collaboration networks. II. Shortest paths, weighted networks, and centrality. Phys. Rev. E **64**(1), 016132 (2001)

22. Nikiporets-Takigawa, G.: Tweeting the Russian protests. Digital Icons: Stud. Russ. Eurasioan Cent. Eur. New Media **9**(2013), 1–25 (2013)

23. Page, L., Brin, S., Rajeev, M., Terry, W.: The pagerank citation ranking: bringing order to the web. Technical report, Stanford University (1998)

24. Vieweg, S., Hughes, A.L., Starbird, K., Palen, L.: Microblogging during two natural hazards events: what twitter may contribute to situational awareness. In: Proceedings of the SIGCHI Conference on Human Factors in Computing Systems, pp. 1079–1088. ACM (2010)

25. Yardi, S., Boyd, D.: Dynamic debates: an analysis of group polarization over time on Twitter. Bull. Sci. Technol. Soc. **30**(5), 316–327 (2010). http://bst.sagepub.com/cgi/doi/10.1177/0270467610380011

Languages of Russia: Using Social Networks to Collect Texts

Irina Krylova, Boris Orekhov, Ekaterina Stepanova, and Lyudmila Zaydelman[✉]

National Research University Higher School of Economics, Moscow, Russia
krylova93@gmail.com, nevmenandr@gmail.com,
stepanovayekaterina@gmail.com, luda.zaidelman@yandex.ru

Abstract. In this paper we outline a method of finding texts in minor languages of Russia in social networks by the example of VKontakte. We find language-specific markers – special tokens that contain letter combinations unique to a certain language and highly frequent in texts in this language. We use Yandex.XML to generate lists of web-pages that contain texts in these languages. We then download data from web-pages in the https://vk.com domain through Vkontakte API.

Keywords: Minor languages · Lexical markers · Social networks

1 Introduction

There are over a hundred national languages in Russia, excluding Russian and languages that are official in other countries. In this paper we use a term "minor languages" despite the fact that some of the languages count more than a million native speakers. However, linguistic tools for all of these languages are equally few. The lack of tools – first and foremost, the corpora – results from the lack of digitized texts in these languages. The Wikipedia seems to be the most obvious source of such texts and it does, in fact, contain sections in many of Russian national languages. But as Orekhov and Reshetnikov [1] showed in their paper, the Wikipedia is rarely (and in case of minor languages – very rarely) a relevant linguistic source of texts. The goal of this project is to collect texts in Russian national languages which will then be used to create text datasets (like marked-up corpora, sets of n-grams and so on).

2 Why We Use Social Networks

Web-pages of regional newspapers and local municipal bodies are quite common on the Internet, and either all or most of the texts found in such web-pages are written in the national language of the respective region. Nevertheless, we chose against using these pages as the main source of texts in favor of the social networks, primarily Russia's most popular one – VKontakte (https://vk.com). We do not discard common web-pages completely, though, as shown later in the paper.

© Springer International Publishing Switzerland 2016
P. Braslavski et al. (Eds.): RuSSIR 2015, CCIS 573, pp. 179–185, 2016.
DOI: 10.1007/978-3-319-41718-9_11

There is a number of reasons behind this decision. Firstly, even though a social network contains pages in all kinds of different languages, these pages are still identical as far as their structure is concerned. For us as developers this means that we are able to create a universal page-processing tool once we understand what the structure of the page is. With usual web-pages, on the contrary, the structure would differ from one web-page to another.

Secondly, social networks often provide an API and this makes page-processing even easier. An API offers a number of methods that allow third party software to access the social network's data. This means that at this point the structure of the pages becomes irrelevant and the problem comes down to using the necessary method.

Finally, it is the social aspect of social networks. They encourage natural, live communication between actual people, and the texts produced in the process are natural as well, as opposed to automatically generated ones often found on the Wikipedia. Pischlöger [2] mentions additional advantages of social networks for minor languages from the point of view of users in that social networks are cheap, easy to use, and provide communication over long distances; he also states that informal use of language in social networks lowers borders that exist in written language, which is important for people without formal education.

3   How It Works

In this section we describe the technical aspects of collecting texts written in the minor languages of Russia.

There are different techniques for gathering web-corpora. For example, Boleda et al. [3] in order to collect Catalan corpus used as initial (seed) list of domains from a Spanish search engine Buscopio and then crawled other web-pages that either had the.es suffix or were assigned by IP to a network located in Spain. After that they used language filtering to separate Catalan from other languages, applied duplicate detection and successfully gathered 166 million word corpora.

The most popular method for large corpora gathering is to use search engine queries to gather seed URLs from this search engine result page and then crawl these URLs. This method called WaC (Web as Corpus) was first proposed by Baroni et al. [4] and is now used by various researchers [5, 6].

In [5] Guevara used this method to collect a Norwegian corpora. He made up a list of frequent words based on a dump of the Norwegian Wikipedia, then took the top 2000 of them and used different pairs of these words to find pages in Norwegian. Finally, by limiting the results to the pages in the no domain, he gathered a list of seed web-pages in Norwegian.

In our work we mostly follow the steps described in [5], limiting the results to the vk.com domain. In general, in our project we search for different sites in national languages of Russia, however this paper describes only the part of work that deals with the most popular Russian social network. Instead of the top frequent words from the Russian Wikipedia, we use the so-called lexical markers. Unacceptability of most frequent words from Wikipedia for collecting corpora of national languages of Russia was proved by Orekhov and Reshetnikov in [1] at the example of Bashkir, Tatar and some other languages. Most frequent words in these languages, according to the respective versions of the Wikipedia, are water-related terms "river" and "basin", which contrasts with the more common idea of function fords being the most frequent in a language. The concept of lexical markers and the process of selecting them is discussed in the next section.

3.1 Lexical Markers

A lexical marker of a language is a word that is unique to the language and therefore uniquely defines it. We use such words to find web-pages (including pages on social networks) that contain texts in Russian national languages. We collect lexical markers manually from grammars, vocabularies and phrasebooks for the languages in question. Obviously, automatic marker search would be preferable but it is currently impossible as explained later.

Our goal is to find as many web-pages as possible so lexical markers need to be frequent in the language. As a result all of the collected markers are function words.

On the other hand, we wanted to avoid finding pages that contain texts in languages other than the one we look for. That is why the markers are required to be graphically unique and not to occur in other languages. We understand, that a marker is unique by posting it to the Yandex search engine and analyzing search result pages. Mostly, it is not that difficult to identify the page language, as most of them contain nation or country name in the title.

Apart from these compulsory restrictions on markers there is an additional one: markers should only contain Cyrillic symbols. In texts found on the Internet symbols containing diacritics are often replaced with their Cyrillic analogues that are either graphically or phonetically similar to them, i.e. Bashkir "ҡ" replaced by "н" or "ө" replaced by "о". This phenomenon is called "everyday written language" [7]. Some of the manually collected markers contain diacritics so it is necessary to provide a way to replace them consistently, though the symbol pairs may vary depending on the language. It is also necessary to check if a marker remains unique after the replacement procedure. This additional restriction can be easily explained: people who speak Russian national languages usually have a Russian keyboard layout and they often forget or just do not want to switch the layout to Udmurt, thus they write without using diacritics.

The combination of these restrictions is the reason why making the marker search automatic is impossible. Obviously, a rather small set of texts would be quite sufficient for determining the most frequent words in any of the languages we are interested in. However, we would also need to make sure that the markers do not match any word in any other language. This task requires far larger collections of texts or at least a method to create such collections. This brings us back to the problem outlined in the introduction: such collections currently do not exist and developing a method to create them is the goal of our project.

The number of markers we were able to find varied for different languages: while for Tabasaran we have six markers that meet all of the restrictions (Tabasaran words for "how many", "if", "someone", "bigger/greater"), for Tatar we only have three (Tatar words for "whole", "then", "again"), two of which contain diacritics. We provide the translations for these words but not the words themselves to minimize the number of documents where the markers would occur in an "artificial" linguistic environment.

3.2   Collecting URLs

When we have a set of lexical markers for a language, the next step is to find the web-pages that contain texts written in this language. Our tool of choice for the task is Yandex.XML – a Yandex service that enables automatic search queries to Yandex search engine [8]. The number of queries we can make is limited, in our case the limit is 1000 queries per day. For every language we send each of its marker in a separate query and by combining the resulting domain lists for every marker we get a list of domains that contain texts in a given language.

The next step is to search web-pages inside a domain. We use Yandex.XML for this as well by sending queries that contain the name of a domain and a marker both corresponding to a given language. Currently we use only web-page lists that we get from queries with domain name set to https://vk.com, but we store all web-page lists and plan to extract texts from them.

We made a decision to work with community pages rather than users' personal pages. The decision is based on our assumption that an average user would use languages other than the national language on his page because he would have friends who do not speak it. On the contrary, a community unites people who share common interests or, more importantly, a common language. For this reason first, manually composed lists of web-pages in the https://vk.com domain only contained URLs to communities. Unfortunately, the majority of the pages in the lists we get from Yandex.XML are URLs of users' personal pages, e.g. 286 personal pages out of 450 total for Tatar.

3.3   Processing a VKontakte Page

When we have a list of VKontakte web-pages, we proceed to extract all necessary information by sending queries to VKontakte API. The end result of page processing is a JSON file that contains the following information:

- file metadata: name of the language, creation date.
- list of posts on the community wall.
- lists of comments for each post.
- information about the author of a post or a comment: user id, first name, last name, gender, date of birth, city; user ids and names can be used for corpora cue mark-up, which is useful for linguistic and sociolinguistic research.

3.4   VKontakte API Limitations

While the use of VKontakte API undoubtedly simplifies and speeds up VKontakte page processing, there are also a number of restrictions. Firstly, there is a 3 requests per second limitation for all of the methods. Secondly, there are method-specific limitations, e.g.

comment and wall post retrieving methods can only return up to 100 comments or posts respectively [9]. These limitations combined with a large number of languages and communities do not allow us to proceed as fast as we would prefer, but we are constantly improving our processing toolset to increase the processing speed.

4   Preliminary Results

We have been working on this project for a relatively short time and it is still far from complete, but we already have some results. We have downloaded a number of communities devoted to or using minor languages of Russia, the table underneath provides detailed statistics:

Table 1. Intermediate results: the number of downloaded communities for each language

Language	Total (found manually)	Downloaded	Total (found automatically)	Downloaded	Overlap
Adyghe			26	5	
Avar			17	7	
Bashkir	135	107	787	In progress	11
Buryat			181	59	
Chuvash			315	90	
Erzya			76	26	
Ingush			270	87	
Kalmyk			182	54	
Karachay-Balkar			42	16	
Khakas	4	3	4	2	1
Komi-Zyrian			89	39	
Lak			2	0	
Mari			161	54	
Udmurt	72	53	769	In progress	33
Tabasaran			1	1	
Tatar			450	138	
Tuvan			444	In progress	

We were originally provided with community lists for three of the languages (Bashkir, Khakas, Udmurt), which were manually composed during previous research. Since then we were able to automatically generate new web-page lists for all languages for which there had markers using our wrapper for Yandex.XML. Table 1 shows that the automatically generated lists for Bashkir, Khakas and Udmurt are larger than manual ones, as one would expect. However, a large part of the generated lists are actually URLs that we would not normally consider. Firstly, there are a lot of URLs for users' personal pages, with which we decided not to work. Secondly, in some cases several URLs correspond to a single community, e.g. URLs for communities and URLs for individual posts in these communities.

This does not explain why the overlap is so small, even though we could expect the manual lists to be subsets in automatically generated ones. The main reason is the method we used to determine the overlap: for each of the three languages we simply calculated the number of URLs found in both lists. Unfortunately this method overlooks cases when different URLs refer to the same web-page, e.g. https://vk.com/public85682520 and https://vk.com/novostibarum. Once we download Bashkir and Udmurt communities we should be able to take this into account and provide more accurate and hopefully higher figures.

There is one other problem that we came across when we studied the data for the downloaded communities and that is identifying the language. Because we use lexical markers to find VKontakte communities, we are certain that a given minor language is used in these communities. We cannot, however, guarantee that other languages like Russian are not used as well. Indeed, we find examples of Russian and a minor language used not just in separate replies in a dialogue but in a single reply: *То самое чувство когда понимаешь, что 1 шырпы менан утты яндырдып ебарден.* In this line found in a Baskir-speaking community we see both code-switching and use of the aforementioned "everyday written language" (word *менэн* written as *менан*). The question is: what is this language? Obviously, we cannot say if it is Russian or Bashkir, the answer in this case, as well as in many other cases, lies somewhere in between. Let's now look at another line found in an Udmurt community: *Стив Джобс – почетной удмурт.* What language is this? Any automatic language identifier would recognize this as written in Russian, even though there is a minor mistake in that an adjective почетной and a noun удмурт do not syntactically agree, which is actually rather common for texts found in social networks. However, from the point of view of Udmurt grammar this line is absolutely correct being an actual Udmurt translation for "Steve Jobs is an honorary Udmurt".

We are definitely not the only ones aware of the code-switching problem. C. Pischlöger provides several examples [10, 11] of Udmurt and Russian switching in VKontakte. He calls this phenomenon "suro-pojo" ("суро-пожо" – "mix" in Udmurt) and states that this mix characterizes the contemporary situation of a living minor language. From the words of his informant, "only foreigners speak clear Udmurt".

5   Conclusion

The approach proposed in the paper has proved to be quite effective. We currently have lexical markers for 97 languages of Russia. These markers were used to generate web-pages lists via Yandex.XML and we have so far collected lists of web-pages in the https://vk.com domain for 32 languages and lists of web-pages in other domains for 18 languages. We have also downloaded (completely or partially) communities related to 17 languages using VKontakte API. These are very early numbers and we expect them to increase as we continue our work and collect larger sets of texts in languages of Russia. We plan to share our corpora with the community as soon as we have got more structured and marked-up data.

Acknowledgements.   We thank Timofey Arkhangelskiy for pointing out difficulties of language identification by the example of Udmurt.

References

1. Orekhov, B.V., Reshetnikov K.Yu.: To the assessment of Wikipedia as a linguistic source (К оценке Википедии как лингвистического источника), Contemporary Russian on the Internet (Современный русский язык в интернете), Moscow, Jazyki slavjanskoy kul'tury, pp. 310–321 (2014)
2. Pischlöger, C.: Besermyan in the internet: social networks as a chance for language maintaining? (Бесермяне в интернете: социальные сети как шанс для сохранения родного языка?), Problems of ethno-cultural interaction in the Ural-Volga region: history and the present (Проблемы этнокультурного взаимодействия в Урало-Поволжье: история и современность), Samara, pp. 216–219 (2013)
3. Boleda, G., Bott, S., Meza, R., et al.: CUCWeb: a Catalan corpus built from the web. In: Proceedings of Second Workshop on the Web as a Corpus at EACL 2006 (2006)
4. Baroni, M., Bernardini, S., Ferraresi, A., Zanchetta, E.: The WaCky wide web: a collection of very large linguistically processed web-crawled corpora. Lang. Resour. Eval. **43**(3), 209–226 (2009)
5. Guevara, E.: NoWaC: a large web-based corpus for Norwegian. In: NAACL HLT 2010 6th Web as Corpus Workshop, pp. 1–7 (2010)
6. Ljubešić, N., Erjavec, T.: hrWaC and slWac: compiling web corpora for Croatian and Slovene. In: Proceedings of 14th International Conference, Pilsen, Czech Republic, pp. 395–402 (2011)
7. Zaliznyak, A.A.: Old Novgorod dialect (Древненовгородский диалект), Moscow, Jazyki slavjanskoy kul'tury (2004)
8. Yandex.XML – https://tech.yandex.ru/xml/
9. VK API – https://vk.com/dev/api_requests
10. Pischlöger, C.: Udmurt and Besermyan languages in social networks (Удмуртский и бесермянский языки в социальных сетях). In: Proceedings of International Science-Practical Conference, Dedicated to 260-Anniversary of V.G. Korolenko Материалы Международной научно-практической конференции, посвященной 260-летнему юбилею В.Г. Короленко.), Glazov, pp. 187–190 (2013)
11. Pischlöger, C. Notes from Murjol underground: super Udmurts in cyberspace (Запис(к)и из Муржол Underground: Super удмурты в Cyberspace). In: Proceedings of IV International Science-Practical Conference "Florov's Readings" (Материалы IV Международной научно-практической конференции "Флоровские чтения"), pp. 56–59. Glazov pedagogical institute, Glazov (2014)

Author Index

Printed in the United States
By Bookmasters